THE BONE MUSEUM

THE BONE MUSEUM

travels in the lost worlds of dinosaurs and birds

Wayne Grady

Four Walls Eight Windows
New York/London

Copyright © 2000 Wayne Grady

Published in the United States by
Four Walls Eight Windows
39 West 14th Street
New York, NY 10011
http://www.4w8w.com

First published by the Penguin Group in Canada in 2000

First U.S. printing October 2001

Library of Congress Cataloging-in-Publication Data:
Grady, Wayne.
The bone museum : travels in the lost worlds
of dinosaurs and birds / by Wayne Grady.
p.   cm.
Includes index.
ISBN: 1-56858-204-8
1. Dinosaurs. 2. Birds, Fossil. 3. Birds—Evolution.
4. Grady, Wayne—Journeys. I. Title.

QE861.4.G72 2001
569.9—dc21   2001033712

Printed in the United States
10   9   8   7   6   5   4   3   2   1

For Matt Cohen, 1942–1999
Not nearly long enough

Acknowledgements

I COULD NOT HAVE written this book without the cooperation of many people. Phil Currie and his colleagues at the Royal Tyrrell Museum of Paleontology in Drumheller, Alberta, have been their usual helpful and friendly selves, in particular those who worked with me on the *Giganotosaurus* and *Albertosaurus* sites in Patagonia and Alberta: Paul Johnston, Mike Getty, Darren Tanke, and Stewart Wright. Similarly, Rodolfo Coria and his staff at the Museo Carmen Funes in Plaza Huincul, Argentina, made my work there not only possible but enjoyable. I also thank Phil's wife, Eva Koppelhus, for her help and guidance in Argentina and Canada. Other members and friends of the various expeditions recounted herein include J.-P. Zonneveldt, Jørn Hurum, Thom Holmes, Adrian Gimenez Hutton, Don Lessem, Valerie Jones, Dave Eberth, Don Brinkman, and Wu Xiao-chun.

I also want to thank Tim Tokaryk for his bookstore and his *T. rex*, as well as Sharon and Peter Butala for their hospitality and for making the Wallace Stegner House available during my stay in Eastend, Saskatchewan.

Closer to home, Jackie Kaiser, my editor at Penguin, provided the encouragement and enthusiasm I needed during the writing and editing of the book, and Mary Adachi managed to preserve the book's idiosyncrasies without missing any of its inconsistencies. Meg Taylor straightened out many of the foreign phrases sprinkled throughout the text: apparently I know less Spanish than even I thought I did.

A portion of this book appeared in an earlier form in *Equinox* magazine, and the short section on William Beebe appeared in *Brick* magazine's collection *Lost Classics*, published by Knopf Canada.

And of course, once again, I thank my wife, Merilyn Simonds, for her patience, wisdom, inspiration, companionship, and love.

Table of Contents

THE BONE MUSEUM

travels in the lost worlds of dinosaurs and birds

PART ONE

Flight Paths

Phil Currie's
Christmas Turkey

To THOSE WHO TRAVEL, all routes are circuitous. My route to Patagonia began and ended in Alberta. I had gone there in January to attend a niece's New Year's Eve wedding in Calgary, when a snow storm closed the airport and a chinook opened the roads north of the city. I decided to rent a car and drive to Drumheller, to see what was new at the Tyrrell Museum and to visit some old friends who worked there, the paleontologists I'd travelled with to China's Gobi Desert in 1990 to research my first book, *The Dinosaur Project*. Writers' routes are circuitous, too.

Drumheller is no Shangri-La in January. A former coal town, it was built deep in the Alberta Badlands, three hundred feet below prairie level because that's where the coal was, although the coal wasn't very good and it soon ran out anyway. Samuel Drumheller, a coal baron from Walla Walla, Washington, opened a mine there in 1918, so the town could have been called Walla Walla and should be grateful. Drumheller built a fine house that is still the finest in town, and then, perhaps not wanting to die so close to hell, moved to Los Angeles, the city of angels, and died there in 1925.

Ironically, dinosaurs saved the town from extinction. In the early 1980s, when the Alberta government was looking for a place to build the Tyrrell Museum, Drumheller's town fathers made a convincing pitch to have it located in their dying town, even though at the time there weren't many dinosaurs in Drumheller either. Now, of course, you can't walk down the main street without being stampeded by herds of concrete reptiles. Even the crustaceans are represented, for the very pillars of the bank building are festooned with stylized architectural trilobites. The local hockey team is called the Drumheller Dinosaurs. Every motel, of which there are now dozens, is called Dinosaur This or Jurassic That. The road from the town to the museum is called the Dinosaur Trail. Drumheller has embraced dinosaurs the way Orlando hugs Mickey Mouse. It might have made more scientific sense to build the museum a few hours south, in Dinosaur Provincial Park, where the dinosaurs actually were, but political decisions don't have to make scientific sense, and the museum, hence the town, attracts half a million visitors a year.

I was one of them, maybe even the first of the year. I had a room in the Jurassic Inn, a two-storeyed motel on gasoline alley surrounded by rolling, snow-covered Cretaceous hills, not a Jurassic rock in sight. I drove into the parking lot under a huge replica of the gates in the movie *Jurassic Park*, minus the iron bars and electronics, and parked under a kind of carport over which I half expected to see the head of a rapacious raptor. The room had two double beds, each one beneath a framed print of a dinosaur, one labelled "Dinosauria Tyrannosaurid" and the other "Dinosauria Brachiosaurid." I slept under the tyrannosaurid. In the morning, before the museum opened, I made coffee in my thoughtfully provided in-room coffee maker and sat at the window looking out at the Badlands mounding above the shopping mall across the highway, admiring the way the sun reddened the eastern slopes of the scurfy coulees. To see a horizon in Drumheller, you have to tilt your head and look up. Directly across from the motel, at the top of a distant coulee, was one of those annoying oil-pumping things called a donkey, which spend the whole day going up and down, up and down, sucking up oil for the pipelines. I don't know why they're called donkeys. They reminded me more of those ridiculous plastic birds that were once popular in family rec rooms and mahogany-lined bars, the kind that perched on the lip of your martini glass and tilted precariously into the cocktail and then stood up again, as if to drink. You could only take it for so long before you ripped it off your glass and threw it away, but in Alberta you have to watch these giant dipping mechanical birds going up and down, for days on end, year after year. Oil, natural gas, coal, Alberta's fossil fuels and, of course, dinosaurs, fossil fuel for the tourist industry.

The first person I ran into at the museum was Dave Eberth. I knew Dave from China; as a sedimentologist, he'd been there to study an area we were digging and to tell us what the place had been like when the dinosaurs died. He was very thorough and conscientious, didn't like to speculate on anything until he'd studied all aspects of it, and even then preferred to wait until he'd checked the literature. We'd stop the Jeep and drop him off on our way to the quarry in the morning, and he would spend the day wandering around the Gobi Desert with his Jacob's staff and his measuring tape while we broiled in some convection-oven valley hunched over a dinosaur quarry in 130-degree heat. When we picked him up on the way back to camp in the evening, he would be darker and dustier and dehydrated and quite glad to see us. In the evenings at our camp, which was a Mongolian sheep-herder's winter quarters that had been built with the bricks from a nearby lamasery destroyed during the Cultural Revolution, he fairly hummed with intensity, as though he had picked up energy from the desert wind that whistled through the low, thorny vegetation and over the scorched sand. Though thoroughly fricasseed by the contemporary climate, in his mind he was working through the landscape as it had been seventy-five million years before, when in place of desert there had been wide, slow rivers weaving through gentle bayous and everglades. In the evenings, as we sat around rehydrating with gin mixed with pear juice, he would conjure up dune sequences and crossbedding, ancient shorelines and fertile riverbeds. Then he would take out his guitar and play a few tunes. He played in a blues band in Drumheller and lived alone in a sparsely furnished house. It was hard, he told me one evening, to maintain a relationship when you were on the road for six or eight months of the year, doing geology.

When I saw him that morning in Drum, he had just returned from a frustrating trip to Germany to look at a Permian exposure that had never been described by a geologist before, even though a team of paleontologists were making strident claims about the fossils they were discovering in it. Dave found this irresponsible. Science, he said, is not about hand-waving and headline-making. Science is about accurate measurements, thought percolated through time as groundwater sifts through layers of sand and rock, not conjectures bubbling up out of nowhere like hotsprings.

We talked about the others who had been to China that year. Phil Currie, who had dreamed up the project and was the Tyrrell's chief vertebrate paleontologist, had been going back and forth, working with Chinese colleagues on the new feathered dinosaurs that were coming out of the northeast corner of the country. Dale Russell had left the National Museum in Ottawa to take up a position somewhere in the States. Don Brinkman had spent more time in the Gobi, pursuing fossil turtles. Paul Johnston had not been back, but was working on a new theory of microinvertebrates. In China, Chang Meeman was no longer head of Beijing's Institute of Vertebrate Paleontology and Paleoanthropology; she had been replaced by Qiu Lianhe, one of the senior scientists who had been in the Gobi with us, a specialist in multituberculate mammals, which in the Late Cretaceous Period were no bigger than salamanders. I remembered crawling over sizzling rock searching for their tiny needle-like teeth with a magnifying glass.

"And you've heard about Li Rong," Dave said.

I hadn't. Li Rong was a paleontologist with the Inner Mongolian Museum in Hohhot, the capital, a highly industrialized city of two million people in the middle of the Gobi Desert, most of them Han Chinese workers shipped north by Chairman Mao in the 1960s, during the famine. I have a photograph taken from the top of the Inner Mongolian Hotel: you can hardly see the next building for coal smoke. Li Rong was in charge of the museum's fossil vertebrate section, and so was our guide when we were out digging in his part of the Gobi. We'd called him Li Rong-Way, because he was always getting us lost. Whenever we came to a place where two roads intersected, Li Rong would hesitate for a fatal fraction of a second and then point in the wrong direction. In the Gobi it didn't matter much, because we'd eventually come back to the same crossroads anyway and we were seldom in a hurry, but Li Rong took his duty as host scientist very seriously, and would get flustered and red in the face every time it became painfully evident that he had got us lost again. Then he would hesitate twice as long at the next intersection. For his sake, because we were very fond of him, we would pray for a simple fork in the road, since a choice of only two possible routes reduced his chances of error, but he usually chose the wrong road anyway. He would end up before an astonished shepherd's yurt, or in some steam-filled village restaurant, arguing loudly over a crumpled map spread out on a table, his shoulders sagging in hopelessness. He wore a floppy straw hat tied tightly under his chin with a ribbon and, even when tramping across the desert, a light, tight, grey business suit made of some indestructible fabric, and black, pointy-toed street shoes. When he wasn't lost he was always smiling, and he never complained about the heat or the wind or the lack of proper signs or the sand in his shoes.

One day, after many false turns and switchbacks, he took us to his most famous discovery, China's largest dinosaur trackway, a forty-kilometre stretch of flat mudstone impressed with thousands of bird-like footprints, some as small as pigeon tracks, others large enough to sit in. He pointed out the metre-wide, webbed prints of giant carnivores, the three-clawed prints of reptilian raptors, the delicate hieroglyphics of the bird-footed hadrosaurs. "Many, many dinosaurs," he exclaimed proudly, using the only two words of English he knew.

What I hadn't heard was that a few years ago, Li Rong had been caught selling dinosaur eggs to a Japanese collector. He was arrested, held in prison for a year, and then released on virtual house arrest, stripped of his job at the museum and given some menial task, a janitor at his apartment building, Dave thought. "A few months ago we heard he was dead," Dave told me. "They told us it was a heart attack, but I strongly suspect he killed himself. The disgrace to his family would have been too much for him."

I remembered the night Li Rong invited us to his apartment for dinner, Dave, Phil, Don, Dale and two or three others. How proud he was of his wife, Shari—a beautiful, dark-skinned Mongolian who, like all Mongolians, traced her roots directly back to Genghis Khan—and of his apartment, a one-bedroom unit in a Soviet-style, concrete highrise. He and Shari had the place almost to themselves, sharing it with only one other family who lived in the living room. We sat at a table set up in the bedroom, two of us sitting on the side of their bed because there weren't enough chairs, while he and Shari brought in platter after platter of food: a baked carp; sliced tomatoes under a mound of white sugar; stir-fried green peppers; soft mushrooms in brown sauce; steamed pigs' ears; and chicken, the real, meaty parts, not just the usual beaks and feet. We were embarrassed, because the meal must have cost him two months' salary, maybe more, and he and his wife hardly ate at all. On the wall above the bed was a two-year-old Western calendar with a photograph of a woman wearing a cream-coloured cashmere sweater much like the one Shari was wearing. As a museum employee, Li Rong had been to the United States and Canada, and had purchased things few private citizens in China could hope to own: a cassette player, a small, portable television. In a glass-fronted display case beside the television was an eight-inch statue of Snow White, painted blue and white, the colours of the Virgin Mary, surrounded by seven miniature Buddhist monks.

"Poor Li Rong," I said to Dave, and he winced. The black market for fossils was becoming more lucrative with every new find, and the temptation must have been strong. Li Rong knew it was wrong to sell those eggs, and what the penalty was for getting caught, but he also knew that we had things he couldn't have. The money he must have seen us spend. Restaurant meals, presents for our wives and our kids. I'd spent a hundred dollars for a roll of silk, others had bought carpets and leather coats. I imagined Li Rong in a duty-free shop in Los Angeles or Vancouver, then later, on the plane, the new cream-coloured cashmere sweater with little pearl buttons in the compartment above his head, finding himself sitting beside a Japanese businessman who was interested in fossils. Li Rong-Way.

ON THE MAIN FLOOR of the Tyrrell I walked through the museum's new Burgess Shale exhibit, in which magnified versions of the shale's bizarre Cambrian fauna are suspended in eerily lit space behind a Plexiglas wall and under a Plexiglas floor. A voice-over narrated the story of life on Earth 500 million years ago, when these creatures existed, much of the information culled from Stephen Jay Gould's book on the subject, *Wonderful Life*. In the reading room later, I picked up a copy of the *Alberta Palaeontological Society Bulletin*, and read an article chiding the museum for swallowing whole Gould's interpretation of the Burgess Shale and in particular for incorporating his contingency theory into the narrative. Gould's theory is that evolution is so contingent on accident, luck, happenstance and serendipity that if life on Earth were a film, and if the film were rewound and played again, the odds against it turning out the same would be astronomical. Humans might not have made it. These dream-like creatures from the Cambrian shale might have become the dominant life forms, and the world might have been filled with their descendants rather than those of the dinosaurs. "Contingency is unrecognizable as a scientific theory," the APS Bulletin chided, "and certainly smacks of being a philosophy, no more, no less." Well, it was an attractive philosophy, I thought, and why shouldn't a science museum smack of philosophy? In Aristotle's day, philosophy, the love of truth, was a science.

In the same issue of the APS Bulletin, Phil reviewed an article by Jennifer Ackerman that had appeared in National Geographic. Ackerman's article, called "Birds Take Wing," reported on the new feathered dinosaurs that had been found in China. Liaonang Province, Phil had written, contained "more specimens relating to the origin of birds than all the world's other sites combined." He thought the site would eventually yield thousands of bird species from the Jurassic, Cretaceous and more recent eras, "possibly representing several simultaneous groups of evolutionary paths." The new material was giving us glimpses of whole new and tangled networks of bird origins, with different families possibly evolving from different dinosaur groups, and sorting it all out was going to take time. "There's nothing neat about bird evolution," Phil concluded.

Down on the main level again, I toured the museum's dinosaur hall where, amid whole skeletons of smaller specimens, the skulls of most of the large carnivorous dinosaurs were mounted on separate plinths and arranged around the room's periphery like the busts of Roman Caesars. I walked around calmly gazing into gaping jaws filled with huge teeth, all aimed at my head. Skulls the size of cows, teeth as big as railway spikes.

The dinosaur world is divided into two huge groups, the meat-eaters and everything else. The meat-eaters are called theropods, a sort of shorthand for the designation Theropoda, which was assigned to all meat-eating dinosaurs in 1881 by Othniel Marsh. The word means "beast foot," and was meant to distinguish the bipedal carnivores from the Sauropodomorpha (sauropod for short, meaning "lizard foot") by the structure of their ankles and feet. *T. rex* is a theropod. So is *Albertosaurus*, slightly more distant in time and slightly smaller than *T. rex*. They were both huge animals, lizards bigger than elephants.[1] Their teeth were pointed, serrated and grooved, made for tearing and gulping, not for chewing; they would clamp onto a victim and simply hold it until it bled to death, then they would rip it apart and swallow it in whole chunks. Their forearms were too short to be of much use around the mouth, but they had strong hands with sharp claws, obviously designed for clasping and holding. Their hind legs were massive drumsticks, gigantically clawed, and they moved quickly: *T. rex* could run twenty-five miles an hour. There was *Sinraptor*, the theropod Phil and others found in China, looking slightly squashed, and beside it, a *Carnotaurus* skull, short, pugnacious snout and horn-like ridges on its forehead, from Argentina. It was a rogue's gallery, no doubt about it. The world's most wanted villains.

After viewing the theropods, I bought one of Phil's books in the museum's gift shop—the weighty *Encyclopedia of Dinosaurs*, which he edited with Kevin Padian—and took it into the cafeteria. But instead of reading it I sat beside the window and watched snow falling on the patio outside. Just beyond the tables was a life-sized sculpture of an *Albertosaurus*, exact in its unknowable details, the pebbled skin, the stern, eagle-like disapproval in its eye. It was looking back into the cafeteria and snarling at me, its ridiculously small

---

1 *T. rex* weighed about seven tons; Jumbo, the biggest elephant in the world, according to P. T. Barnum, who owned him and ought to have known, weighed six tons.

arms seeming to wiggle in helpless fury. As I watched, a magpie landed on its neck. No matter how often I see magpies, they startle me. This probably marks me as an Easterner in the eyes of many Albertans, who consider magpies to be more of a nuisance than a boon. But I get the same pleasure from blue jays, to which magpies are related. It has something to do with pure colour, I think. Magpies are pure black and pure white, and the lines dividing the colours are sharp and clean. This magpie—the name means pied Maggie, or speckled Margaret, and is very old: we get the word "pie" from the bird, not the other way around—stood on the *Albertosaurus's* neck and looked about to see what mischief it could cause. Magpies are, according to all the IQ tests that have been devised, the smartest of birds, possibly the smartest of animals. Next to us, of course.[2]

I hesitate to call myself a birder because I know too many real birders, but I do keep a life list, I always carry binoculars and bird books with me when I travel, and I take the time to use them, so I guess I qualify. I became interested in birds during a visit to Point Pelee, a narrow finger of land hooking down out of southwestern Ontario into Lake Erie, famous as a resting stop for birds at the intersection of two major migration routes. My parents used to take me swimming there when I was five. I went back a few years ago in early May, when the shorebirds were resting up and the warblers were coming through. It was as though I had never seen a bird before, and now I was bombarded with them, twenty, thirty, forty species in a single day. Since then I have seen many birds, but I have never forgotten that sudden jolt into awareness, and I experience it again every time I see a new bird, or renew an acquaintance with an old one. Like this magpie. There seems to be no end to the wonder they excite in me.

"Theropods," Phil wrote in the *Encyclopedia*, which lay open on the table before me, "first appeared in the Triassic and survived

2 In British folklore, magpies are counted to foretell the future. The counting rhyme is: "One for sorrow, two for mirth, three for a wedding, four for a birth. Five for rich, six for poor, seven for a witch, I can tell you no more." The last line is obviously polite for: "Eight for a whore."

more than 160 million years. In fact, because birds are the direct descendants of theropods, the theropod lineage is still very success-ful today and has a history of 230 million years."

I looked out at the magpie again, and thought: it's sitting on a statue of its ancestor. Here was a whole new source of wonder.

I DROPPED IN TO SEE Phil the next afternoon. He was in his lab, a white room connected with his corner office that looked like a medieval reliquary or ossuary, it was so full of bones. The bones were old, 80 million, 100 million years. I still find it impossible to imagine such distances, anything lasting that long and still looking like itself. I walked around, looking down at the counters and tables and up at cupboards and shelves, all scattered, heaped, with petri-fied bones. A fossil pterosaur, one curved finger of each hand longer than its whole body, as though it had died holding two canes. The spine of a juvenile ankylosaur. A cast of the *Albertosaurus* skull I'd seen downstairs in the dinosaur hall, which had been found by Joseph Tyrrell himself in 1884. The pelvis of a *Ceratosaurus*, another of the large theropods. A *Tyrannosaurus rex* skull from South Dakota perched on a windowsill; beside it, possibly for com-parison, possibly for companionship, rested another huge theropod skull, *Daspletosaurus*, gazing out at the Alberta Badlands from which it came.

Among these were the contentious specimens, the cross-dressers of the Cretaceous, the smaller theropods that were half dinosaur and half bird. No one was suggesting that *Tyrannosaurus rex* evolved into a bird; birds were already around long before *T. rex* appeared in the Late Cretaceous Period. But there were many earlier, smaller theropods, turkey- to ostrich-sized creatures distantly related to *T. rex*, that exhibited bird-like characteristics, and these were Phil's special interest. A cast of a *Mononykus* arm and claw, identified first as a dinosaur but which Phil now said was probably a bird. A flat cardboard box contained parts of an *Ornithomimus* skeleton, the ostrich-mimic, a true theropod and yet toothless like a bird. The box was marked: "Delicate, Do Not Disturb." There were also a number of less exotic items, skeletons of actual birds: two ostriches, a gull, a crow. Phil isn't eccentric enough to refuse to call them birds, but he thinks of them as modern dinosaurs. "Pluck the feathers off a bird," he has said, "and you've got a dinosaur."

The previous June, at the annual meeting of the Society of Vertebrate Paleontology, Phil delivered a paper that announced one of the most significant scientific discoveries of the century: he and Mark Norell and two Chinese colleagues, Ji Qiang and Ji Shu-An, announced that a pair of dinosaurs, which Phil and his co-authors named *Caudipteryx* and *Protarchaeopteryx*, had been found in northeastern China with distinct feather impressions surrounding the skeletons. They were not birds, they were dinosaurs, clearly theropods, but parts of their bodies, their arms and tails at least, had been covered with feathers. The announcement stunned the assembled scientists. Dave Eberth, who'd been at the meeting, told me that Phil's announcement was followed by a prolonged silence as two hundred minds struggled to assimilate the information. "You could almost hear the paradigm shifting," he said. Then came the applause. For Phil, and for many of the scientists in the room, dinosaurs with feathers provided the final, dramatic proof that modern birds are directly descended from theropods. "Although both theropods have feathers," he told the gathering, "they are both more primitive than the earliest known bird, *Archaeopteryx*. These new fossils," he concluded, "represent stages in the evolution of birds from feathered, ground-living, bipedal dinosaurs." They were a kind of missing link, tangible elements in a line of descent that, until these miraculous beings turned up, existed only in the theorists' imagination, a so-called ghost lineage. The story, complete with rebuttals from rival theorists, was covered in *Time* and

Maclean's, the *New York Times* and CNN. The discovery and Phil's involvement with it changed his life. It was more than the fulfillment of a prediction; it was a stunning and dramatic conclusion to years of controversy, the final summation to a scientific debate that had started nearly a hundred years before Phil was born.

He was sitting at the end of a cluttered table, carefully removing bones from a crumpled paper bag and laying them out on paper towels. A week before, the bones had been part of his family's Christmas turkey. After dinner, Phil had salvaged the carcass and boiled it on the kitchen stove, then picked the bones from the bottom of the pot and laid them out on the kitchen counter to dry. Today he'd brought them into the lab for reassembly. One of science's big attractions is that you get to play with things. Who doesn't like to play with bones?

Phil is tall, heron-like, with the arms and gait of a baseball pitcher—in fact, he reminded me of Bill Lee, who pitched for the Boston Red Sox and whom I met in Montreal when he was finishing off his career with the Expos. Like Lee, Phil seemed to be a knot of concentrated calm at the centre of a raging world, part of it and apart from it at the same time. I remembered working beside him in the desert, in an ankylosaur quarry, when a sandstorm arose, wind and rain blowing so suddenly and intensely that the very air seemed to have become stinging vapour; Phil calmly pulled on a pair of ski goggles and continued working. The rest of us followed his lead, and eventually the wind stopped. He was doing what he loved. He has loved doing it since he was six years old, when he found a plastic dinosaur in a box of cereal. When he was eleven he told his parents he was going to be a paleontologist. His father worked in a can-making factory in Oakville, Ontario, and didn't know what a paleontologist was, so he said, Sure. Phil remembers the dinosaur. "It was a *Dimetrodon*," he says, "and it was in a Rice Krispies box. The Rice Krispies people in the States say they never put dinosaurs in cereal boxes, but maybe the Canadian Rice Krispies people did. They must have, because all I ate were Rice Krispies, and I didn't like them very much. I only ate them to get the dinosaurs." He still had the plastic *Dimetrodon*. He took me into his office next door and showed it to me. He also showed me the first dinosaur book he ever read: Roy Chapman Andrews's *All About Dinosaurs*. He read it when he was ten. In it, Andrews wrote of the excitement of digging for dinosaurs in Mongolia, of being chased by bandits one moment and beset by sandstorms the next. He also had a copy of *The Wonderful World of Prehistoric Animals*, by William Elgin Swinton. He handed that to me, too: it was signed: "To Philip Currie, With the Compliments of W. E. Swinton, Aug. 1962." In 1962, Phil was thirteen.

"I visited Bill Swinton when he was a biologist at the Royal Ontario Museum in Toronto," Phil told me. "He was fantastic. He had written a lot about dinosaurs and was very enthusiastic, had lots of great stories and good advice for a kid of my age. He told me not to take biology in high school, for example, but to take math and physics and wait until I got to university for biology. He didn't tell me there were no jobs in paleontology. He just said if paleontology was what I wanted to do, then I should do it."

For the next hour or so Phil kept taking down books and handing them to me. One of the Audubon Nature Series, with illustrations by Charles Knight on perforated stamps: someone, young Phil, had pasted the coloured illustrations over black-and-white squares. "That was 1959," he said. A copy of *Life* magazine from 1952, with the cover-lines: "The World We Live In, Part 2: The Earth Is Born." Inside were reproductions of the famous Zallenger murals from the Peabody Museum, in Yale University, showing (usually swampy) environments from various distant eras of Earth's history, with every known animal of the time standing around looking self-consciously at the painter, like a group portrait of politicians at an unpopular convention. "A bit crowded, maybe," said Phil, "but I was fascinated by these pictures."

Early on in his pursuit of fossils he realized that dinosaurs did not come from Ontario, that no matter how assiduously he prospected along Sixteen-Mile Creek, behind Oakville, he would never find anything more exciting than clams and cephalopods. "I collected those," he said, "cleaned them and preserved them and labelled them. It was good practice, but it wasn't the same." When he was twelve, he persuaded his father to take him and two of his three brothers west for their summer vacation. They drove through the Dakotas and Wyoming and Montana, stopping at places where great discoveries had been made, "humouring me," Phil said, "so I could see where the dinosaurs came from." They must have viewed most of these places from a distance, the Badlands of North Dakota from the roadside, the fossil forest in Yellowstone from a campsite, because Phil saw no actual bones until the night they stopped at a motel in Alzada, Montana. While he and his brother unpacked, Phil's father went for a drink in the bar, "a real saloon," Phil recalled, "with swinging doors and everything." After an hour, his father came back and told Phil to come with him. The two of them went back into the saloon and the bartender reached under the counter and took out a box of bones. "He stuck them on top of the bar and told me I could go through them. They weren't dinosaur bones, I could tell that even at that age. I didn't know what they were, but the bartender told me I could pick any one I wanted and take it home, so I did. When I got home I identified it as a mosasaur bone.[3] I still have it."

Also when they got back to Oakville, Phil realized that most of the dinosaurs he'd wanted to see came from western Canada, not

---

3 Mosasaurs were huge marine creatures, and finding their bones in Montana is a reminder that all of central North America was once seabed, the floor of the vast inland saltwater sea that once connected the Arctic Ocean with the Gulf of Mexico. During most of the Cretaceous Period, when the Alberta dinosaurs roamed, the Badlands were coastal.

from Montana or Wyoming. On his next trip to the Royal Ontario Museum, after his visit with Swinton, he looked at the labels on the exhibits and saw that his favourite dinosaurs had been found in the Alberta Badlands. "I hadn't been smart enough to read the labels before," he said. "It was kind of ironic, given where I've ended up."

"THE BUTCHER CUT OFF the head and part of the neck," Phil said, piecing the turkey together along the tabletop. The bird's backbone lay stretched out between us like the skeleton of a snake. Exactly like the skeleton of a snake, in fact, for as T. H. Huxley observed in 1866, surprising even Darwin, chickens were "an extremely modified and aberrant Reptilian type," little more than snakes with feet, and both the chicken and the snake it swallowed belonged to the same "province," which he called the Sauropsida. The first salvo in what turned out to be the bird-dinosaur wars was fired by Huxley more than 130 years ago.

"He also cut off the feet," Phil added, meaning the butcher, "which is really too bad. But we do have a nice vertebral column." He picked up a turkey vertebra and placed it on the table beside one from a *Dromaeosaurus*, a small theropod from the Badlands. Except for their size, they were almost identical. "If you didn't know one was fossil and the other modern," he said, "you'd almost swear they came from the same animal."

In 1976, before finishing his Ph.D. at McGill, Phil took a job at the Provincial Museum in Edmonton and set about fulfilling his dream of prospecting in the Badlands. Gradually, he filled the museum with dinosaur material almost literally from its own backyard. In Montreal he had been studying a group of Permian reptiles called eosuchians, small, slithery life forms that may have given rise to snakes and lizards. Phil was examining amphibious members of the group known as phytosaurs, the ancestors of modern crocodiles. In those days, he said, "everyone was championing the idea that birds came from crocodiles, and I felt that there was a fair bit of evidence to suggest that that was true. But I didn't think about it very much." The notion that they came from dinosaurs, as proposed by Huxley a hundred years before, had died out in the 1920s with the appearance in English of Gerhard Heilmann's *The Origin of Birds*.[4] Heilmann pointed out many similarities between theropods and birds—the scaly legs, the long necks, the fused foot bones—but concluded, almost regretfully, that theropods lacked the one vital component that would clinch the argument: clavicles, or collarbones, which in birds eventually fused together to form the wishbone, or furcula. The first bird known to science, a 140-million-year-old fossil named *Archaeopteryx*, had a wishbone. Dinosaurs had once had clavicles, though unfused ones, but they seemed to have evolved out long before *Archaeopteryx* came on the scene, and the rule of evolution is, once a feature is lost it is not regained. Once we humans evolve out our appendices, for example, which we appear to be doing, we will not evolve them back again. It's the same with dinosaurs and collarbones. And without clavicles there could be no furcula, and without a furcula no animal could fly. Case, according to Heilmann, closed. *Archaeopteryx* must have evolved from something else, thecodonts for example, which were early reptiles that evolved into crocodiles and had kept their clavicles. After Heilmann, the dinosaurian origin of birds was a non-issue for the next fifty years.

In 1973, however, the year after Phil began working on his doctorate, an American paleontologist from Yale University, John Ostrom, published his comparisons between *Archaeopteryx* and a small theropod he discovered in Montana, just south of the Alberta border. Ostrom named his new dinosaur *Deinonychus*, which meant "terrible claw," because the little beast had huge, scimitar-shaped

---

4 Heilmann's book was published in Danish in 1917, but went virtually unnoticed until it was translated into English ten years later.

claws on its feet that were sharp enough to slice the jugular of its prey or enemy with a single kick. In his 1973 paper in *Nature*, Ostrom revived Huxley's contention that birds must be descended from dinosaurs. The similarities between *Archaeopteryx* and *Deinonychus* were startling; moreover, since *Deinonychus* had to have been very agile to use those claws effectively, and in order to be that athletic it probably had to be warm-blooded, *Archaeopteryx* had probably been warm-blooded too. And if the first birds were warm-blooded, they couldn't have been descended from the sluggish, cold-blooded thecodonts, clavicles or no clavicles.

At that time, Phil wasn't particularly interested in bird origins. "I had no tendency either way," he said, though he admitted that if pressed on the matter he probably would have sided with the crocodiles. Then Larry Martin paid a visit to the Alberta Provincial Museum.

Martin was an avid defender of the thecodont origin of birds: he contended that birds split off from the evolutionary tree way back in the Early Triassic Period, at the very beginning of the age of dinosaurs. They might have had a common ancestor with dinosaurs, but then everything on Earth has a common ancestor if you go back far enough. A professor at the University of Kansas, Martin is still happy in his role as "the Antichrist of the bird-dinosaur theory," as he calls himself. A portly, animated, avuncular favourite with graduate students and fellow researchers, he oozes common sense. I've seen him stand before an assembly of scientists and evoke an image of *Archaeopteryx*, with small claws at the ends of its winged arms, trying to run along the ground like a dinosaur, grasping at a fleeing animal with its long flight feathers getting in the way, and of course it was clearly impossible. He was lecturing to a room full of committed bird-dinosaur defenders, at John Ostrom's own university, at a conference dedicated to John Ostrom's dinosaur-bird hypothesis. Ostrom himself was sitting in the front row, while a display of feathered dinosaurs from China was being set up in the Peabody Museum next door, and Martin was bellowing like King Canute against the rising tide of bird-origin specialists: "Function studies," he said, wagging his finger at them, "looking at a piece of anatomical equipment and trying to determine logically what it was used for, is far more useful than phylogenetics," which he characterized as the bloodless configuration of dinosaur categories according to some computerized calculation of composite characters. "Are these configurations defended by great science," he asked rhetorically, "or great scientists? Why *would* a dinosaur put claws way out at the ends of its wings?" *Archaeopteryx* was not a dinosaur, never was. So what if it had claws: a clear case of convergent evolution. It did not inherit claws from its dinosaurian ancestors. It grew them on its own, to aid in climbing trees. "You'd be suspicious if you saw a Kentucky Fried Chicken wing with little fingers on it, wouldn't you?" he said. Yes, yes we would. "Well, you should also be suspicious when you see them on *Archaeopteryx*."[5]

How ironic, then, that it was Martin who got Phil thinking seriously about the dinosaur origin of birds. Phil had borrowed a *Dromaeosaurus* from the American Museum of Natural History, and while he was showing Martin around the Alberta Museum, Martin pointed to that specimen to illustrate how dinosaurs could not possibly be ancestral to birds. All dinosaurs, he told Phil, had small bones in their jaws that held their teeth in place, called interdental plates. He pointed to the plates in the *Dromaeosaurus* specimen. Phil looked, and they were indeed there, "fused together so you could hardly see them, but quite clearly there." Birds and crocodiles do not have interdental plates, Martin said, and therefore birds are more closely related to crocodiles than they are to dinosaurs.

---

5 After Martin's performance at the conference, I heard one paleontologist at the coffee counter say to another: "You know your theory is in trouble when you have to resort to common sense to prove it."

"After Larry left, I looked into it more closely," Phil said. "I found he was right about *Dromaeosaurus*, but I also had a *Troodon*"—another small theropod from Alberta—"and I looked at it, and guess what? No interdental plates. Larry also said that in both birds and crocodiles, but not in dinosaurs, you find a constriction in the tooth between the crown and the root. Well, *Troodon* had that constriction as well. In fact, the more I looked at this little dinosaur the more bird-like it became. Then, when we finally found a *Troodon* braincase in Dinosaur Provincial Park, I realized that a lot of things were telling me that dinosaurs were very closely related to birds, and probably ancestral to them."

This is jumping ahead a fair bit, because the *Troodon* braincase wasn't found in Dinosaur Provincial Park until 1986. By that time the defenders of the bird-crocodile hypothesis were in retreat. In the end, Martin was overtaken by the very common sense he advocated, for you need only look at a *Troodon* skeleton, with its hinged ankle, its clawed, three-toed feet, its long, powerful legs, short, grasping arms and its gracefully curved neck, to see its resemblance to a bird. Except for its long tail, it might be an ostrich, or a swan. Sure, *Troodon* had teeth, but other small theropods didn't.

There were other similarities, too, less obvious to me. Phil held up the fused vertebrae from the tail of his Christmas turkey and showed me a photograph of an almost identical structure from a new species of oviraptorid dinosaur from Mongolia, called Nomingia. The fusion of vertebrae at the end of the tail, he said, "is probably a mechanical development meant to help support feathers at the end of the tail." You see them most clearly in the two specimens from China, he added, in *Caudipteryx* and *Protarchaeopteryx*. Tails were particularly interesting. Most theropods had long, tapering tails, whereas the tails of modern birds are short. But bird tails are in fact long tails that have been balled up, the vertebrae fused into a fistlike knot called *pygostyle*. The tail of Phil's turkey had twenty-two vertebrae, the same number as in the tail of Nomingia and Caudipteryx, and the tail feathers of the latter radiated in a fan from the pygostyle, very closely resembling a turkey's tail. As he said, the more Phil looked at dinosaurs the more they resembled birds. And even I could see that the more we looked at birds, the more they resembled dinosaurs.

For someone interested in birds, this was an amazing discovery. Could I add *Archaeopteryx* to my life list, I wondered? If I touched a bone from *Hesperornis*, would it count as a sighting? But for those concerned with the larger picture, with trying to discern a unifying structure beneath the complexities and diversities of the view of nature that science has given us, it was even more compelling. For Carolus Linnaeus, classifying the Earth's myriad species in the eighteenth century, Nature was order, if only we could discern it. Disorder was Chaos, in Milton's phrase, ". . . a dark,/ Illimitable ocean without bound,/ Without dimension, where length, breadth and highth,/ And time and place are lost." For us, it is Nature that seems chaotic, disconnected, illimitable, a billion things happening independently and at the same time. This view has largely been Darwin's legacy, and it has led Western society along some destructive pathways: if nature is disconnected, for example, we can destroy part of it without threatening the whole. We suspect this isn't the case, we have a kind of species memory of it not being the case. In fact, we desperately need to know it isn't. Living in such a fragmented universe is insanity. Every now and then a great discovery comes along that gives us a glimpse into the connectedness between seemingly disparate things, a clue that tends to confirm our suspicion that everything in nature is linked to everything else. Someone demonstrates that the behaviour of the Humboldt Current off the coast of Chile, for example, can cause an increase in hurricanes along the eastern seaboard of the United States, or that controlling insects with pesticides in Indonesia can poison the breast milk of a polar bear on Wrangell Island. Most of the ex-amples of interconnectedness seem to be negative, tending to point out to us the degree to which we have inadvertently messed up the planet. But I began to see the connection between dinosaurs and birds as one of those discoveries, and it was relatively benign. We hadn't screwed up; it had happened 150 million years ago.

But paleontology doesn't escape controversy simply because it is concerned with the distant past. Throughout the nineteenth century, and for much of the twentieth, paleontology was accused of breaking up the comfortable picture of the world that religion once provided for us. It was paleontology, for example, especially the discovery of *Archaeopteryx*, that provided support for Darwin's wildly conjectural and decidedly anti-Christian hypothesis that human beings were descended from apes. Paleontology had intro-duced the concept of extinction into the world: species appeared, flourished, peaked, crashed and disappeared. This was not consid-ered good news. This was not what we had come to think of as God's plan. Could it be that, by showing how two apparently unconnected things in nature—hummingbirds and *Tyrannosaurus rex*, for example—are in fact intimately linked, paleontology could put some of that comforting picture back together again? Maybe extinction wasn't so final, after all.

ON MY WAY OUT, Phil took me down to the preparation lab on the main floor of the museum, not the large prep room with a glass wall through which bored museum-goers watch museum staff work-ing on yet another hadrosaur from the Badlands, but a smaller room in the remote back of the building in which greater reckonings were at play. Jim McCabe, one of the Tyrrell's preparators and another China alumnus, was delicately freeing the tiny, fossilized skeleton of a bird from a slab of grey rock in which it had been encased for the past 120 million years.

"*Confuciusornis sanctus,*" Phil said, leaning over the slab. Jim had been at it for a month already, and figured he'd need another month to complete the job of preparing it.

Its name means "the sacred bird of Confucius."[6] It was hardly the size of a magpie, and lay splayed out on the rock as though prepared for dissection, or as though it had been sunbathing on its back when something big, a boulder or a sauropod's foot, had flattened it like a Cretaceous roadkill. Then several metres of lake mud and volcanic ash had settled on it, which then compressed slowly to rock, encasing the specimen forever. Its legs and wings were spread open, there was a fan of feather impressions where its tail was, and the skull at the end of its long neck was turned slightly to three-quarter profile, so that its big orbitals and toothless beak were presented for easy inspection.

"It was found by a farmer in northeastern China," Phil said. "There are about a thousand *Confuciusornis* specimens now. Only about three hundred are available for study, however. The rest have been spirited away to private collectors, sold on the black market. We traded two, legally, from the National Geological Museum in Beijing."

I asked Phil if he would be going back to China. "Not this year," he said. "There are so many people looking for feathered dinosaurs up there now. Most of them farmers and unskilled as paleontologists, but still, I'd have to be incredibly lucky to accomplish much in the short time I'd be able to spend there. All the obvious places have been thoroughly prospected."

---

6 In Chinese mythology, Meng Niao, the country of birds, was in present-day northern China, "north of Mo." An early manuscript relates that the people of Meng Niao were skilled at taming birds, but the later, tenth-century *Treatise on Nature*, purporting to be more scientific in its description, says that the beings in this country were actually bird-human hybrids, with human heads and bird bodies. Their feathers were speckled with red, yellow and green. At some point in their history, the bird-people rebelled against their Hsia overlords and fled south, a mass movement that was looked upon so favourably by the gods that birds of paradise were sent to accompany them. The treatise was meant to explain not only migration, but also how humans first began to eat eggs.

One of the less obvious places that had not been picked over by hundreds of amateur or even professional paleontologists, he said, was Argentina. In his book *The Flying Dinosaurs*, Phil described a chicken-sized bird with long legs and short arms, evidently a flight-less runner (no wishbone) from the Late Cretaceous beds of north-western Patagonia. It was the seventh fossil bird to have been discovered in Argentina since 1991. (In fact, eighty per cent of all the fossil bird species known to science have been discovered in the past decade.) "Superficially," Phil had written, "it seems to have changed little from its theropod ancestors, although it was in real-ity a descendant of a flying bird species." In other words, it seemed to have taken an evolutionary path later followed by ostriches. Argentina was emerging as a focal point for theropod and bird dispersal.

"Whereabouts in Argentina?" I asked him.

"Northern Patagonia," he said lightly, but there was no denying the thrill that the word Patagonia injected into the air. Patagonia was the end of the Earth, the farthest reach of the known world. Patagonia. I thought, of course, of Chatwin and Theroux, of Darwin and W. H. Hudson, all of whom had begun their careers in Patago-nia. "We opened a dinosaur quarry there last year, and a bunch of us are going back again in March to take out the rest of the bones."

"A bird?"

"No, a huge theropod, bigger than *T. rex.*, though there are birds and bird-like theropods all around there. Patagonia has become the theropod capital of the world," he said. "If ten new dinosaurs are found each year, four or five of them will be from Patagonia, and most of those will be theropods. If you want to study theropods, then Patagonia is definitely the place to pitch your tent."

The quarry Phil was going to work in was way out on the steppes, he said, inaccessible except by four-wheel drive. "We're working with an Argentine paleontologist named Rodolfo Coria;

he'll pick us up in a little town called Neuquén and drive us in to the quarry. Takes about two hours. It's his turf, of course. We'll camp there with him for a month or six weeks, depending on how much help we have. There are some fascinating aspects of this new dinosaur, sort of related to the bird-dinosaur hypothesis. It has bird-like characteristics. It's totally unrelated to any North American dinosaur species, and yet evolutionarily it is very close to them. It'll be fun to see just how close."

"What's it called?"

"So far it's unnamed, but it seems to be related to a dinosaur called *Giganotosaurus*, which at the moment is the largest theropod in the world. This one might be bigger."

"Patagonia, you say?"

It was odd, standing there in one of the most modern of the world's natural history museums, looking at a fossil no bigger than a pigeon, and talking about the world's largest dinosaur from one of the world's remotest regions. Two hours from Neuquén, I thought. Where the hell was Neuquén? I felt a familiar tug, the traveller's itch, that curious, irresistible downslope into the other. Patagonia.

"Need any help?" I asked him.

Phil hesitated for a second. "Sure," he said. "We'll need all the help we can get. You'd be welcome." Then he paused again, and I knew what he was thinking.

"I'll pay my own way," I said, having no idea how I would do that.

"We'll have to charge you for food," he said. "And you'll have to pack in all your gear."

"No problem."

"It'll be like China," he said. I didn't know whether he meant it as encouragement or as a warning. I took it as both.

CHAPTER TWO

Lost Worlds

I WISH I COULD SAY that science has always held a special fascination for me, that as a child I exhibited a precocious understanding of the secret workings of the physical universe. But it hasn't and I didn't. So many of the scientists I have met have told me that their interest in science began at an impossibly early age, practically *ab ovo*, later to be nurtured by (as in Phil's case) a plastic dinosaur from a cereal box, or a friend's Meccano set or a caterpillar slowly asphyxiating in a Mason jar. Science, they have told me, is a particular way of looking at the world, a kind of physiological predisposition, like a second language, difficult to acquire after puberty. One geologist I know began collecting rocks and minerals when he was six. He became so ardent at it that his parents gave him a special room in the house for his collection, and when he eventually left home to go to university he had ten thousand specimens, which he carefully packed and took with him. A biologist told me that when he was ten he began picking up roadkills on his way home from school, car-struck cougars and the like, and boiling them on the kitchen stove to get at their bones. He was sent off to a boarding school at an early age. A paleontologist friend has a nice collection of animal and bird craniums, catalogued, labelled and mounted, that he has been accumulating for as long as he can remember. He still smuggles skulls back from cycling tours in Latin

America; I once saw him rolling a pelican skull up in a blanket, and was glad we were leaving the country by different routes.

Such was not my fate. For me, the butterfly in the Mason jar never emerged from its chrysalis. I grew up in Windsor, Ontario, a mean, hard-nosed city across the border from Detroit, although by that I do not mean to blame the city. The plastic toys that came in cereal boxes in Windsor were usually cars, and I didn't like cars much, either. I suppose I collected those nature cards from packages of Red Rose tea, like everyone else, with their murky paintings of hoary marmots and three-toed sloths, but my interest in them, as with hockey cards later on, was merely acquisitory. I remember my Aunt Joan in Newfoundland sending me stacks of Audubon Nature cards at Christmas; I admired them briefly, but they were hardly fertilizer for a nascent career in zoology, and while there may have been dinosaurs on some of the cards, as well as golden tamarinds and stoats and other equally shaggy and mysterious creatures, they do not, alas, stand out in my memory. A duck-billed platypus from Australia, a duck-billed hadrosaur from Alberta, would all have been the same to me.

So far as I know, there was no zoo in Windsor, or if there was I did not go to it. No Museum of Natural History. I don't even remember there being a public library, although I'm sure there must have been one. I do remember piles of foundry sand in the welder's yard next to our house, where the frameworks of huge transport trailers, the kind used to haul automobiles to dealers' lots, would be parked waiting to be repaired and towed away, looking for all the world (come to think of it) like mounted dinosaur skeletons. In the piles of sand we would find large chunks of very fine sandstone, so soft we could break it in our small hands and shape it by rubbing it on the sidewalk, or by carving it with pocket knives. It wasn't the sandstone that fascinated me. I did not say, Hey, this is Geology and it's fun! I carved the blocks into toys. Mostly ships, as I recall. I would rub them on the concrete until they had blunt prows and skewed funnels, then I would push them abstractedly through the red sand, pretending the foundry slag was an ocean and I was crossing it. But even this exploratory urge was buried under layers of obscurity. When Miss Finch, my grade-three teacher, had us all stand up, one after the other, and announce what we would be when we grew up, I did not say: I'm going to be an explorer. Possibly thinking of the ships rather than of their destinations, or of the men who welded things in the shop where the sandstone came from, I said: "Mechanical engineer."

"Oh no," Miss Finch said, shaking her straight red hair. "Not an engineer; you're going to be a writer."

"Oh," I said, and sat down.

Though I was not a child of science in Windsor, I must have been one of nature, because when nature did occur to me I seem to have been ready for it. I eventually discovered that even a city like Windsor had animals and trees and birds. There was a field across the street from our house. Well, we called it a field. It was big enough to have a railway track running through it, with a ditch on either side, and grass and brush growing wildly about, and I discovered that where there is a ditch there will be tadpoles, and where there are tadpoles there will be snakes and birds. I was in Windsor a few years ago and, walking nostalgically through the old neighbourhood, I saw that what we had called a field was in reality little more than a vacant industrial lot. But I grew up in the days when city councillors could look at a vacant lot and, if they couldn't think of anything intelligent to do with it, they just left it alone. Now I suppose they'd pave it and put a swing set in it and one of those homicidal merry-go-rounds, call it a parkette and name it after someone they owed money to. But then it was my first wilderness area. It teemed with wildlife. I explored it daily. I crawled through the grass, squiggling under the tall shedding heads, inch by inch, partly to evade detection by the troop of enemy soldiers I was convinced had been hiding out in a secret underground bomb shelter since the war, but partly, mostly, to see what kind of creatures lived there, besides myself. I would flatten a few square feet of grass, put down some collapsed cardboard boxes for a floor, and make a series of pathways, undetectable from the air under their canopy of grass. The pathways radiated out from this central fort, stretching to the far corners of the known field. Crawling along these tunnels, with green sunlight filtering through their grassy roofs, I encountered a perplexing diversity of urban wildlife, garter snakes, birds, insects, groundhogs, semi-feral cats. No golden tamarinds or crested cockatoos, but who knew what was around the next bend? The field was my introduction to nature. Crawling through it on my stomach was my initiation to field work.

AFTER THAT, MY NEW interest in nature might have been fanned by the first movie I saw as a sentient being, which was *King Kong*. My father had bought our first television set, a large, square, polished wooden box that sat on a large, square, polished wooden table. It was the most beautiful piece of furniture we owned. It glowed in its corner even before it was turned on. The day he brought it home, my father told me *King Kong* was going to be on that afternoon, and then sent me to the corner store to buy some pop and potato chips. When I got back, my mother had already pulled the heavy chintz curtains in the living room and Dad had switched on the television set (I think this was before its name was abbreviated to TV, but why did we call that singular thing a "set"?) and they were watching it. It was a blue light in a darkened corner, and when you gazed into it you were drawn through its portal into another world. It was the opposite of Alice's looking-glass, in which Alice encountered familiar objects reversed. Television takes a totally bizarre situation and makes it seem perfectly natural. Then it moves you along to yet more bizarre situations.

Take *King Kong*. You have a giant gorilla climbing up the Empire State Building, carrying a woman in the palm of one hand while being strafed by an out-of-work WWI fighter pilot, and you find yourself saying, Oh yeah, that could happen. In this respect, even a movie seen on television differed from the same movie seen in a theatre. You expect extraordinary things to happen in theatres—that's what theatres are for—and so they do. But television happens in our living rooms. Only ordinary things can happen in living rooms. Me swallowing a filbert and it getting caught in my throat, and my cousin Dumpy holding me by the legs and banging my head on the floor until I coughed it out. That was real life. Television took a giant gorilla terrorizing New York, and made it seem ordinary.

So it wasn't the actor in the XXXL gorilla suit at the top of the Empire State building that impressed me about *King Kong*. Even at the age of seven, I could see that that part was faked—the gorilla's jerky, stop-action movements, the bits of building falling in slow motion even though Fay Wray's arms and legs were wriggling a mile a minute. The part of the movie that held me, that drew me into a world I accepted as real, that I could dream of visiting myself one day, was the part when Carl Denham went to the lost island in the East Indies to find King Kong in the first place and film it. That part wasn't fiction, it was documentary. Gorillas lived in the East Indies, didn't they? And one could get that big, couldn't it? Why not? And if they captured it and brought it back to North America to show it off, well, it stood to reason that the thing would break loose and rampage the city. Happened every day. The rest was just inevitability working itself out.

When King Kong arrived in America the best part of the movie was over for me. I wanted to stay back on the island, where the men in cloth caps and proto-Tilley jackets hacked their way through the jungle, defending themselves from all manner of strange creatures.

I didn't remember until I saw the film again as an adult that the strange creatures they were defending themselves from were dinosaurs. First they're attacked by a *Stegosaurus*; then their raft is overturned and some of them are gobbled by an impossibly carnivorous *Brontosaurus*; then Kong saves Fay Wray from being chomped by a *Tyrannosaurus rex* and carried off by a predatory pterodactyl. I suppose a budding paleontologist would have noted those details, but I didn't.

I PREFERRED BOOKS to movies. When I was nine, we moved out of Windsor to a small farming village in central Ontario, and there I read a lot more books than I saw movies. There wasn't a theatre for miles, but the Bookmobile pulled into our schoolyard once a week. After school I was surrounded by the country, which touched chords in me that reverberate to this day, and, Walt Disney notwithstanding, nature and literature are still more intricately linked in my consciousness than nature and movies. The bush began behind our house and went on as far as my imagination would allow, which is to say forever. I was a city boy let loose in a vast, unexplored wilderness, and when I went out into it I usually had a book with me. First there was a huge stand of tall cedars, and I wandered in its limitless wastes until I knew them as intimately as I had known the field across our old street in Windsor. When I had explored it thoroughly, identified the house sparrows and the crows, counted the frogs, numbered the insects, ferreted out the damp, exposed roots of the cedars, many of which had been blown over by Hurricane Hazel, I had but to cross a horizonless expanse of prairie grassland, negotiate an indifferent herd of grazing buffalo (pretending not to see their ear tags), and enter a larger bush, a proper forest of hills and hardwoods. Here were new animals to identify—raccoons, porcupines, an occasional deer—and new trees to climb—giant oaks, scarlet maples, butternuts and ash. Turbulent rivers tumbled through raging ravines and fed into vast lakes that, by tying fence rails together with bailer twine to make a raft, I could pole across to distant shores, foreign farms, uncharted woodlots, pastures slowly reverting to wildflowers and bush. In winter I could track the mighty jackrabbit to its lair of brambles, then snare it, skin it, cook it and devour it while warming myself on a sun-soaked rock, where it invariably tasted of peanut butter and jam. Onward, ever onward. The thrump of grouse in mating season, the finely pointed print of a fawn in wet snow, a red squirrel's outraged staccato from a nearby pine, the chamois-soft paper of a book read in dappled sunlight, the first spring breeze riffling the pages as I dozed.

The book might have been *Tarzan of the Apes*, which I thought of as the King Kong story told from the gorilla's point of view: what is Tarzan (as he puts it himself) but an ape with a superior intellect? He didn't get trapped and hauled off to America, he went to America of his own accord and was trapped when he got there. When it was published in 1914 it was an instant success, and since then has been almost constantly in print, and has been turned into comic books, movies, several television serials and an animated film. Edgar Rice Burroughs, its Chicago-born author, spun out a whole library shelf of sequels, in all of which Tarzan is a man of superior strength, intelligence and compassion, Mowgli with muscles. Paul Theroux, in his essay "Tarzan Is an Expatriot," describes Tarzan as "a wise metamorph, powerful, gentle and humourless. The animals all knew him," and adds that "he was undeniably a man and bore not the slightest trace of simian genes." This description, based on Theroux's early reading of the comic-book version, would, I think, surprise Burroughs. Theroux was trying to paint Tarzan as the other, the non-belonger, someone who, like Theroux himself, was in the jungle but not of it. In Theroux's memory, Tarzan "asserted his authority over the animals very passively," and "never bit or clawed any of his enemies." Good Victorian that he was, Tarzan was never quite sure whether he was an ape or a human. That's the point of the novel—the issue of Tarzan's genetic makeup is kept deliberately cloudy.

The plot is simple enough. In the 1880s (the height of the Victorian era) a young couple, the woman pregnant, is set ashore by a mutinous crew of cutthroats on the west coast of Africa. They are John and Alice Clayton, the English Lord and Lady Greystoke, who had been on their way to do something administrative in one of the colonies. Lady Alice delivers her son, and about six months later she and her husband die—a kind of failed Swiss Family Robinson. Their infant son is taken by a female ape and raised by an ape clan. He thinks he's an ape, too, but an odd one: small, white and hairless. When he is about seven, he finds the cabin in which his parents died; among their effects is a knife and an English-language primer, and he sets about learning how to read and write, though not to speak, English. It takes him about fifteen years, but he doesn't have much else to do.

Just about the time Tarzan is lip-reading his way through *The Decline and Fall of the Roman Empire* or some other light Victorian tome, another small group of whites is set ashore by another mutinous crew of cutthroats, at exactly the same spot as the first. This second group is made up of an American professor and his assistant, the professor's nineteen-year-old daughter, Jane, her black maid, Esmeralda, and a young Britisher, Cecil Clayton, the son of Lord Greystoke, who is the younger brother of the original Lord Greystoke who died in the same cabin this new group discovers and happily occupies (after burying the old bones they find therein). Clayton is of course in love with Jane, and she with him, though they are both too civilized to mention it. Jane's father is Archimedes Q. Porter, of Baltimore, a professor, I think of natural history, although Burroughs never quite says what he is a professor of. He sees a lion and, instead of running, calls it by its Latin name,

Felis, and that is supposed to stand for all learnedness. He and his assistant, Samuel T. Philander, are portrayed as absent-minded, short-sighted, amiable buffoons who seem to be in the jungle merely to provide a plausible excuse for Jane's being there. About Esmeralda the less said the better. Why is it that so few intelligent people at the turn of the century could discuss Darwinian evolution and inherited characteristics without being racists?

In *Tarzan of the Apes*, Burroughs sets up a social genetic experiment. What would happen, he is asking, if a human being with civilized genes were plucked from civilization, never in fact knew that civilization existed, and was placed in a wild environment? Would he grow up civilized or brutalized? Which would win out, heredity or environment? In fact, he says, let's make it more interesting, let's not even let him know he's human. Then we'll stand back and watch the fireworks. Will blue blood out, or will *Homo sapiens* revert to type?

For a while it looks as though Tarzan's simian genes will win. He conquers his enemies not passively, not by "attitude," as Theroux remembers it, but by brute force: he kills his stronger ape rivals with a combination of quickness, cunning and the knife. When the others arrive, Tarzan is all set to take over as the new King of the Jungle. Then Tarzan sees Jane. Then Tarzan picks Jane up and swings off into the jungle with her, and we all think we know what's going to happen. Even Jane thinks she knows, and is not all that opposed to the idea: "A feeling of dreamy peacefulness stole over Jane Porter as she sank down upon the grass where Tarzan had placed her, and as she looked up at his great figure towering above her, there was added a strange sense of perfect security." When Tarzan makes a nest for them for the night, she experiences a certain frisson of virginal uncertainty, but decides to "trust in fate." Fate, it turns out, is not on her side, for Tarzan is suddenly gripped by his genetic makeup: "... in every fibre of his being, heredity spoke louder than training." Rather than treat her as apes treat their mates, Tarzan suddenly, to his own and Jane's consternation, turns considerate and gentlemanly. Burroughs's prose becomes laden with pseudo-scientific jargon. Tarzan's gestures are "the hallmark of his aristocratic birth, the natural outcropping of many generations of fine breeding, an hereditary instinct of graciousness which a lifetime of uncouth and savage training and environment could not eradicate." In the morning, Jane's virginity is intact, and Tarzan's civilized, British genes have successfully shouldered their way to the frontal lobes of his oversized brain.

Burroughs was in fact examining one of the central issues of Victorian science, one that the Victorians themselves had not resolved. In 1868, in his essay "Man's Relations to the Lower Animals," T. H. Huxley announced that science had established "beyond all doubt the structural unity of man with the rest of the animal world, and more particularly and closely with the apes." But not even Huxley, not even in the 1880s, would have gone so far as to suggest, as Burroughs does, that an ape could mate with a human being and produce a viable offspring: "My mother was an ape," Tarzan says (in French) to the new Lord Greystoke, and the new Lord Greystoke believes him. "No one," Huxley wrote, "is more strongly convinced than I am of the vastness of the gulf between civilized man and the brutes; or is more certain that whether *from* them or not, he is assuredly not *of* them." But Tarzan *thought* he was of them. Here we have the curious split in the Victorian attitude towards evolution. In order for Man to evolve from Apes, they thought it must have been necessary, at some point in history, for an ape to have had sex with a human being. They thought this was what Darwin and Huxley were proposing. This notion did not sit well in Victorian England. The Victorian mind, mired as it was in bizarre notions about human sexuality, got the mechanism of evolution wrong and, having got it wrong, decided not to think about it. Human beings were above that sort of thing. Burroughs, although writing in 1914, was reassuring his readers that one could be descended from an ape, could be raised by a colony of apes, could even think he was an ape, and still behave like an English gentleman.

I didn't glean all of this at the time, of course. *Tarzan*, like *King Kong*, simply whetted my appetite for adventure and nature: I was all for following Professor Porter into the jungle for whatever unspecified scientific purpose, and was disappointed when he turned out to be a mere treasure seeker. After that, I thought everyone in the book was an absolute idiot, including Tarzan. The identical crews of mutinous cutthroats, the foppish Cecil Clayton, the stock vaudeville Esmeralda, *oh mah Lawd*! Burroughs had plainly never set foot in Africa, never met a British aristocrat, never laid eyes on a pirate, nor listened carefully to the speech of a black woman from New England. Like Theroux, I fantasized about Jane and that bower for a few nights, and thought Tarzan a bit thick to leave her panting. And when he killed a lion on a bet (the comic books don't mention that part) and left the jungle to board a ship for America to be with Jane, well, we are suddenly back in *King Kong* country, aren't we.

ANOTHER BOOK THAT may have fallen from my lap in that dappled clearing was Arthur Conan Doyle's *The Lost World*. I came to this scientific extravaganza after breezing my way through the complete Sherlock Holmes. It's hard to say why Conan Doyle killed off Holmes in favour of writing "boys' books," as he called them, like *The Lost World* and the other Professor Challenger novels, but he had always been interested in science, not all of it forensic, and paleontology was as speculative (and increasingly as popular) as detective work. In many ways, it *was* detective work. Scientific investigation could turn up anything imaginable, and some things that were not. On his wedding trip to Greece, Conan Doyle convinced himself that he had seen an icthyosaur swimming in the Aegean Sea (it was almost certainly a dolphin, since icthyosaurs had been extinct for several million years), and later, in his study, along with the personal trophies all writers accumulate—theatre programmes, stream-polished pebbles, old sports equipment—he kept an *Iguanodon* bone that he had found in a quarry near Crowborough, and a cast of an *Iguanodon* footprint from Sussex. Science, as Tennyson observed[1], works on the slow layering of time, the impossible past superimposed on an unlikely present to suggest an improbable future. Perhaps Conan Doyle tired of his insufferable detective and his idiotic companion, Dr. Watson, as I had, and longed to break out of the fogged-in alleys of Victorian London and see the multitextured world.

The Lost World also deals with a pressing issue of Victorian science—evolution and extinction. It was published in 1912 (after being serialized in the *New York Herald*, the same paper that, a few decades earlier, had sent Stanley to Africa to find Livingstone, another expedition I would have killed to go on), and is a novel of exploration, scientific in its scaffolding and romantic in its framework. A movie was made of it in 1925, but I haven't seen it. Ray Bradbury saw it, though, and it changed his life: "The dinosaurs that fell off the cliff in *The Lost World*," he wrote in 1982, "landed squarely on me. Squashed magnificently flat, breathless for love, I floundered to my toy typewriter and spent the rest of my life dying of that unrequited love." Sadly, in the book, it is not dinosaurs that fall off the cliff, but Indians and scientists, thrown over the edge by malicious ape-men. Michael Crichton stole the title (he might call it honouring the memory) for his sequel to *Jurassic Park*, but there is more of Conan Doyle's sense of wonder in Crichton's first book

---

1 In "Locksley Hall": "Science moves, but slowly, slowly, / creeping on from point to point."

than there is in the second, which is more about fear, an emotion related to wonder (the awesome-awful twinset) but much less interesting. And Conan Doyle at least got the time right: although Crichton called his novel *Jurassic Park*, most of the dinosaurs he placed in it were Cretaceous. When Conan Doyle imagined a plateau in South America untouched by evolution since the Jurassic Period, he very sensibly put Jurassic dinosaurs on it. Like Kong's island, Conan Doyle's Maple White Land (named for a poet and artist from Detroit who had died discovering it) had to be a hard place to get to, otherwise the dinosaurs would have got out and might still be walking around Oxfordshire. But it couldn't be so remote that his heroes couldn't get to it. South America was a deeper and darker place than Africa, seventy per cent of it being uncharted. It was still a brave New World and had such creatures— reversions, sports of nature, living fossils—lurking in it. Conan Doyle filled it with a strange conglomeration of science and gobbledygook: dinosaurs and porcupines, pterodactyls and modern birds, ape-men and humans.

This mixture of the extinct and the extant is the most interesting thing about *The Lost World*, scientifically speaking. Perhaps because there was no sex in it, no evolution seems to have taken place either. The dinosaurs isolated on that precipitous plateau had been up there for 140 million years, and they had not evolved at all. It is as though the human travellers had gone back in time to see the dinosaurs, rather than that the dinosaurs had progressed in time to meet the humans. Although one of Conan Doyle's scientists asserts that "evolution is not a spent force, but one still working," you'd never know it from looking at his dinosaurs.

A few years ago, paleontologist Dale Russell, then with the Canadian Museum of Natural Science, conducted a thought experiment; using his knowledge of evolutionary processes, he imagined what a small theropod dinosaur, a *Troodon*, would look like today if it had been allowed to evolve over the past sixty-five million years. He came up with a vaguely alien-looking creature that walked on two legs, had lost its tail, swung its forearms like a salesman who has just landed a major account, and gazed about its new world through huge, frog-like eyes that bulged yellowly from their reptilian sockets. It looked a bit like Jar-Jar, the lizard-like creature in the latest *Star Wars* sequence, only smaller and more intelligent. Conan Doyle's dinosaurs, by way of contrast, seem not so much lost in time as stuck in it. Here is his description of a group of five herbivorous *Iguanodons*, for example: "In size they were enormous. Even the babies were as big as elephants, while the two large ones were far beyond all creatures I have ever seen. They had slate-coloured skin, which was scaled like a lizard's and shimmered where the sun shone upon it. All five were sitting up, balancing themselves upon their broad, powerful tails and their huge three-toed hind-feet, while with their small five-fingered front-feet they pulled down the branches upon which they browsed. I do not know that I can bring their appearance home to you better than by saying that they looked like monstrous kangaroos, twenty feet in length, and with skins like black crocodiles."

This tallies pretty well with what scientists at the time of Conan Doyle's writing thought *Iguanodon* must have looked like. In 1902, the brilliant Belgian zoologist Louis Dollo exhibited his reconstruction of a group of *Iguanodons* that had been discovered twenty years earlier in a coal mine near Brussels: whereas British scientists had depicted *Iguanodon* as a giant lizard or a sort of reptilian rhinoceros—fat, squat, quadruped and stupid—the Belgian exhibit showed them as bipedal, graceful, even delicately constructed, something between a bird and a mammal. Indeed, as models for his revolutionary conception of *Iguanodon*, Dollo had used a flightless bird and a marsupial: an emu and a wallaby. Conan Doyle was up on his science, and like Burroughs he was raising a serious question about evolution. There shouldn't have been

Iguanodons in Maple White Land at all; they should have evolved out, given way to something else, something more advanced. Birds, maybe. Conan Doyle seems to have created a lost world in which there was no extinction. An un-Darwinian realm. But in doing so, he ended up demonstrating that without extinction there could also be no evolution.

This may explain why, although dinosaurs are the excuse for Professor Challenger's expedition to *The Lost World*, they play a fairly insignificant role in the action once the scientists get there. Filmmakers, those ringmasters of the twentieth century, may have pushed dinosaurs onto centre stage (or over the cliff) to impress the cheap seats, but Conan Doyle was more interested in the conflict between the ape-men and the modern Indians who somehow peopled the plateau. He does not suggest that anything from the Jurassic Period actually evolved into people: he rather uses them to explain that there must be a way up to the plateau somewhere that his explorers didn't know about. But the focus is nonetheless on human development, where *Homo sapiens* came from, and where it might be going. As in *Tarzan of the Apes*, the conceit of *The Lost World* is the ascent of man, but where Burroughs suggested that primitive apes and more modern creatures interbred to produce human beings, Conan Doyle's guess is that there was, at some point in prehistory, a clash between apes and men, in which the newer, smarter, gregarious, proto-European sub-group wiped out the primitive, war-like, languageless apes. There is a great battle between the ape-men and the Indians, which the Indians win (with decisive help from the Brits, who have rifles). At the moment of victory, Professor Challenger seems to forget that he is on an isolated plateau in modern South America, and exults as though he had actually travelled back in time and influenced the course of history (perhaps Conan Doyle's original idea had been to put his Londoners into some kind of Wellsian time machine): "'We have been privileged,' he cried, strutting about like a gamecock, 'to be present at one of the typical decisive battles of history—the battles which have determined the fate of the world.'" In fact that Darwinian battle had been fought some thirty-five thousand years before Professor Challenger was born; the war, if there was one, was long over; the battle in which Challenger and his colleagues participated could only have been a fascinating but meaningless skirmish in a remote corner of the world to which news of peace had yet to be delivered.

But Conan Doyle is doing something clever. He is saying in effect that evolution does not work in slow, gradual changes from generation to generation, and species do not come into new being by interbreeding and assimilating and adapting, as Darwin imagined it. In Conan Doyle's world, a species takes over a territory by killing off whatever was there before. A superior species wipes out an inferior one, and all is right with the world. Dinosaurs did not die out but remained the superior species. Human beings evolved because they somehow discovered language and gunpowder. This is close to what Huxley was getting at when he mused on the possibility that nature proceeded by jumps rather than by slow, creeping adjustment to environmental change or species interaction, as had been suggested by Darwin.

THE VICTORIAN'S ATTITUDE towards time was as ambivalent as their attitude towards sexuality, and contributed to their false idea of how evolution worked. On the one hand, they viewed Time as an old friend, and became obsessed with the measurement of it: Greenwich Mean Time, established in 1884, made such things as time zones possible. Time made them able to catch trains and, in factories and collieries, to calculate profits. Before paleontology became fashionable, however, and for a long while afterwards, they had little notion of deep time, the yawning stretches of what

Shakespeare called "the dark backward and abysm of time." Few intelligent people still believed in Archbishop Ussher's assertion that God had begun creating the Earth at nine o'clock in the morning on Monday, October 26, in the year 4004 B.C. (a figure arrived at by adding up the ages of the various generations listed in the Begatitudes), or that dinosaurs perished because they were too big to get on the Ark. But most, if pressed, would have put the age of the planet at about a million years, and the age of the dinosaurs a few thousand or so ago. That some dinosaurs could have survived on an isolated plateau in South America for a few thousand years was plausible, just as it was that ape-men could still exist somewhere. In Africa, probably. Even those who accepted the hypothesis that Man was from the apes seem to have taken the idea of descent quite literally, believing that human beings were descended from apes in the same way that I am descended from my great-grandfather. And since I and my grandfather co-exist, so could apes and human beings. This was part of that fuzzy thinking about sex. They believed that, in evolutionary terms, blacks represented a transitional form between apes and whites: Africa was thus a lost continent, because apes, blacks and whites all co-existed—and interbred—there, just as dinosaurs and modern birds co-existed in Conan Doyle's lost world. When Bishop Wilberforce caustically asked Huxley whether his ancestral ape was on his grandfather's or grandmother's side, he was only half joking. In the Victorian mind, evolution was a kind of extended family tree. If one could trace one's ancestry back to the time of William the Conqueror, one might as well go a few generations farther to include the apes. Except that Victorians did not want to do that. When the wife of the Bishop of Worcester heard that Darwin was proposing that human beings were descended from apes, she is said to have remarked: "Let us hope it is not true, but if it is, let us hope it does not become generally known."

It is this deliberately miasmic thinking about evolution that John Fowles depicts in *The French Lieutenant's Woman*, set in England in the 1860s. When Fowles's main character, an amateur paleontologist, discovers that he and the village doctor share a common passion for Darwin, they feel as though they belong to a secret society of progressive thinkers: "They knew they were like two grains of yeast in a sea of lethargic dough—two grains of salt in a vast tureen of insipid broth." Imagine how Darwin himself must have felt.

The Victorian view of evolution ignored Darwin's greatest breakthrough, which was the identification of natural selection as the mechanism that drives evolution. Even Huxley balked at it. If natural selection could produce a new species, he failed to see how it was that artificial selection, as practised by pigeon fanciers and sheep breeders for centuries, had not produced a new species of domestic pigeon or sheep. Darwin himself admitted that it had not. All domestic dogs, for example, with their multitudinous racial variations, were yet a single species: a Mexican chihuahua could still mate with a German shepherd and produce pups. If natural selection created new species, then chihuahuas and German shepherds ought to be different species by now. Huxley therefore accepted Darwin's hypothesis only provisionally: "So long as all the animals and plants certainly produced by selective breeding from a common stock are fertile, and their progeny are fertile, and their progeny are fertile with one another, that link will be wanting." If so dedicated a Darwinian as T. H. Huxley could entertain doubts about evolution, then perhaps it was possible (though unthinkable) that a British peer could mate with an ape and produce Tarzan.

THE DEBATE OVER THE origin of humankind was paralleled by another, similar argument, one that though overshadowed by it was no less far-reaching in its implications. That was over the origin of birds. The origin of human beings and of birds is one debate, for if human beings could be descended from apes, it followed that anything could be descended from anything: horses from hippos, hawks from handsaws. Birds could be descended from fish, or insects. Huxley suggested reptiles. It was like the word game in which you must change one word into another by changing only one letter at a time; thus APE becomes MAN by the series APE, APT, OPT, OAT, MAT, MAN. Huxley was saying it was possible to get from CROC to BIRD.[2] Reject the latter and you are bound to reject the former; but accept the former and you cannot logically reject the latter. Therefore, Victorians rejected the idea that birds were descended from reptiles. "Not long ago," Darwin wrote in *The Descent of Man*, which, when it was published in 1871, upped the ante on the outcome of the debate, "paleontologists maintained that the whole class of birds came suddenly into existence during the Eocene period," that is, long after the extinction of the dinosaurs. The discovery of *Archaeopteryx* in 1861 rendered that belief untenable. *Archaeopteryx* was clearly a bird and lived at the same time as the dinosaurs. "Hardly any recent discovery," Darwin wrote, "shows more forcibly than this how little we as yet know of the former inhabitants of the world." Although it took a while for its full implications to be realized and accepted, paleontology had changed forever the way Victorians viewed the world. Scientists like Lord Kelvin who busied themselves calculating the age of the Earth based on the rate at which fresh water became salty, or the temperature of rock at the bottom of mine shafts, had thought dinosaurs roamed the Earth as recently as a hundred thousand years ago. But for Darwin's theory to work, for one species of dinosaur to evolve into another, millions upon millions of years, unimaginable gulfs of eons, would be necessary; the mind boggled at the prospect of dinosaurs evolving into birds. Just as the Copernican revolution in

---

2 You can do it, but it's not pretty: CROC, CROW, CREW, BREW, BRED, BLED, FLED, FEED, FEND, FIND, BIND, BIRD.

astronomy changed forever our concept of Space, Darwin's hypothesis and its insistence upon gradual change irretrievably altered our awareness of Time.

I HAVE BY NOW ACQUIRED an interest in dinosaurs, but I am much more interested in why so many people find dinosaurs interesting. To scientists, part of their attraction is that they represent a possible complete set of information: the fossil record is notoriously scanty, but the animals are extinct, and so at some possible future date, may be completely known. The link with birds makes them even more intriguing to me, and evolutionists can use them to try to piece together the stages in the journey from one to the other. As paleontologist Jack Horner has written, "now that we know [dinosaurs] were less reptilian than avian they seem closer to us," the way, he suggests, that modern space technology makes us feel closer to aliens. Dinosaurs "are the only large-scale aliens that we can be certain visited the planet." But Horner acknowledges that the true lure of dinosaurs is the mystery of their existence, the mystery of life itself, "the very fact that there is life here at all and that everything that's alive today is so because everything else passed away." This is a profound, very nearly a religious, observation, that life is made possible by death, that evolution needs extinction. In order to understand life, we need to understand extinction.

What did the Victorians know about extinction? Only thirty-four years before the ascent of Queen Victoria, Thomas Jefferson, president of the American Philosophical Society as well as of the United States, expressed his unswerving opinion that, given the remarkable economy he had observed in Nature, "no instance can be produced of her having permitted any one race of her animals to become extinct." This from a man who had caused an exhibit of mastodon bones, discovered in a saltpetre mine in Virginia, to be mounted in the White House. It is very probable that Jefferson believed that the Lewis and Clark expedition into the unexplored regions of the American West would encounter a living specimen, hiding out in a remote, hidden valley in the Rockies, and bring it back with them.

Despite paleontology's mounting evidence to the contrary, a similar belief lurked in the back of the Victorian mind. There was, for instance, Conan Doyle's insistence that he had seen an icthyosaur. Before and during Darwin's day the most widely known theorist in evolution was the Chevalier de Lamarck, who held that there was no such thing as extinction, no old species died out naturally, but rather slowly changed into new species—thus there were always exactly as many species on Earth as there had been on October 26, 4004 B.C. Richard Owen, Britain's most famous paleontologist at midcentury, was a stout opponent of the notion of evolution. To Huxley's audience, the idea that a species could become extinct was horrifying. They looked about them, at the beauty and peacefulness of nature, at the tall smokestacks of industry that spoke so eloquently of progress. Surely, they thought, all this could not end. Though they lived in an age steeped in the impedementa of death, engaged in wars of conquest and extermination that claimed hundreds of thousands of lives, they liked to think that, at home at least, nothing really died.

Darwin and Huxley tried to strip them of that illusion. Nature was not an extension of the Christian, moral universe, they said. Things in nature are killed, they die, and human beings are part of nature. "Nature," Darwin wrote, "may be compared to a surface on which rest ten thousand sharp wedges, touching each other and driven inwards by incessant blows." This was not a Wordsworthian sentiment. After Darwin, Victorians could no longer deny that extinction existed, but they could prefer not to think about it. They could, by supreme mental effort, lift themselves out of the general flux of Nature, shut their minds to the teeth and the claws, and convince themselves that while evolution, as a constant drive towards improvement, happened at home, extinction, evolution's unsavoury corollary, was something that happened in the colonies. And there were places in the world where neither evolution nor extinction existed.

This subliminal belief lasted well into this century, which is one way to explain the popularity and prestige of such Edwardian science-fiction writers as Conan Doyle, H. G. Wells and Jules Verne, all of whom featured dinosaurs in their mythical lost worlds. And it even persists today, when otherwise rational people believe that Bigfoot, Sasquatch and the Abominable Snowman are surviving pockets of Neanderthals who retreated into the mountains thirty-five thousand years ago and mastered the art of astral projection, or that the Loch Ness Monster is, shades of Conan Doyle, a plesiosaur. Michael Crichton knew what he was doing when he honoured Conan Doyle's memory by borrowing his title: we are as willing as our forebears were to believe in the redemption of lost worlds. Even our own.

Which raises the question: What do *we* know about extinction?

A paleontologist friend wears a T-shirt that says: "Extinction Happens." No kidding. Just ask the dinosaurs. All three major geological periods in which dinosaurs flourished—the Triassic, the Jurassic and the Cretaceous—ended with huge mass-extinction events. The Cretaceous came to a crashing halt sixty-five million years ago when a meteorite smashed into the Yucatan Peninsula. The impact wiped out not only all the remaining dinosaurs but more than seventy per cent of life on the planet. This seems definitive. For years we looked upon the end of the Cretaceous Period, which we called the Cretaceous/Tertiary (K/T) boundary, as a barrier, a door slammed on death row, a catastrophe that practically denuded the surface of the Earth, like a huge chalkbrush wiping the slate of Earth's surface almost clean of species.

Lately we've revised that picture. It's true that none of the big dinosaurs survived, but we now think they'd been dying out for a

million years or so before the Big Wallop killed the persistent strag-
glers. There are very few places in the world where dinosaur fossils
are found within a metre of the K/T boundary, and a metre of rock
can represent a million years of deposition. Paleontologists are
beginning to view the K/T more as a filter than a wall, because in
fact a lot of smaller things made it through. Among invertebrates,
most crustaceans made it (only one bivalve disappeared). Shrimps
and lobsters and lots of insects, including mosquitoes, termites and
cockroaches, continued. A great many vertebrates also survived.
Sharks were hard hit but a few species squeaked through, and bony
fish made it through swimmingly. Crocodiles, turtles, lizards, salamanders, frogs, all present and accounted for. Mammals, marsupial
and placental, here. And, of course, birds. And if you call birds
avian dinosaurs, then dinosaurs made it, too. Certain members of
each group disappeared, but enough survived to keep the *families*
going. In fact, just about every family of life that existed before the
K/T existed after it. Perhaps the K/T mass extinction wasn't so massive after all?

We are, in other words, in the process of redefining extinction,
softening its blow. Why? The fact remains that seven out of every
ten living organisms on the planet disappeared. Why are we trying
to make that seem less like a catastrophe and more like, well, like a
culling?

It isn't that we are denying death; we are embracing it. It's as
though in order to prove to ourselves how far we've come from the
prissy Victorians, how inured we are to reality, we've made death so
much a part of life that we regard extinction as a kind of apotheosis. The twenty-first century has already been called "the century of
biology": if, in the nineteenth century, we learned how to count,
and in the twentieth we learned what to count, now, in the twenty-
first, we will add up the numbers to find how short we are from a sat-
isfactory tally. They will tell us that, in our chess match with Death,

Death is up by more than a few pawns. Current estimates are that the world extinction rate is one thousand times greater than historic levels. Peter Raven, director of the Botanical Gardens in St. Louis, Missouri, and co-inventor, with Paul Ehrlich, of the concept of co-evolution, has predicted that a quarter of Earth's species will become extinct in the first twenty-five years of this century. Unless dramatic action is taken to reverse the current rate of habitat destruction, he says, two-thirds of the Earth's three hundred thousand plant species will be lost in the next hundred years. In North and South America alone, sixty-five species of birds are endangered, and hundreds more are threatened. Nineteen species of reptiles, the descendants of those that survived the Cretaceous Period, are on the verge of extinction now. Forty-three species of mammals. Fifty-five species of fish. Recently, Harvard biologist E. O. Wilson estimated that, worldwide, we are losing an average of twenty-seven thousand species a year: that's three species every hour. Some scientists say Wilson's estimate is wildly conservative.

This realization, for some, paralyzes thought and mocks any notion of creativity. Graeme Gibson, a Canadian novelist and an internationally recognized birder, is one. In his novel *Gentleman Death*, the central character, who is a novelist, sits at his desk day after day with the extinction tally ringing in his head, immobilized by the absurdity of trying to write in a dying world. "In the time it takes me to find my two hundred words," he thinks, "over a hundred species of plants and animals become extinct. In case you're interested, that's thirty-six thousand a year. Which is to say for every two words of this, my deathless bloody prose, an irreplaceable form of life vanishes forever. It's a sickening thought."

Here's another: the rate at which species are currently becoming extinct approaches, if it does not exceed, the extinction rate at the end of the Cretaceous Period. We may be living through the fourth mass-extinction event in the half-billion-year history of life on this planet. And that's a thought that, at the societal level, we don't seem to be doing much about. Like the Victorians, but in a slightly more sinister way, we just don't want to have to think about extinction. But whether we think about it or not, we are allowing it to happen. We still think we are "outside nature," exempt from the laws that govern plants and insects and lesser species. It's as though we have come under the grip of some kind of torpid, special death wish of our own.

PART TWO

Patagonia

CHAPTER THREE

To the Rio Negro

NEUQUÉN IS THE CAPITAL of Neuquén Province, the centre of Argentina's oil-producing district in the northern end of
Patagonia. The Air Austral flight from Buenos Aires took about
two hours. I spent part of the time practising my Spanish on the in-
flight newspaper. I don't know why I persist in thinking that I can
read newspapers in a language I barely know and still get the drift
of the story, but I do. I tried it in China with risible effect, and I
attempted it once in Norway, perhaps because I was there to attend
a conference on translation. I was billeted with a Norwegian trans-
lator in Oslo, a very jovial host named Thorstein Haverstad, and
when we walked into town and were having a drink at Ibsen's
favourite hotel, the Grand, I sat there puzzling out an article on the
front page of the *Dog's Bladder*, as I think the local newspaper was
called. After much diligent effort, I determined that the article was
about the imminent defeat of Norway's current prime minister, but
when I asked Thorstein about it he took the newspaper from me,
glanced at the front page, and said the article was announcing the
return of Norway's Olympic athletes from Seoul.

I gave up on the newspaper and took up a book I'd bought in
Buenos Aires: W. H. Hudson's memoir of Patagonia, *Far Away and
Long Ago*, written in 1918 when Hudson was seventy-seven, a poor,
sick, lonely man in a boarding house in England. Every birder

knows Hudson's work. He was born in Quilmes, a small town just outside Buenos Aires, in 1841, his parents having emigrated from New England, where his father had worked in a brewery. Quilmes was a beer-brewing town, although it had once been the site of a reservation set aside for a now-extinct nation of aboriginal people. Hudson describes his early life as a wild, dark and painful childhood spent running free on the pampas, his habitual sorrow and loneliness relieved by his love of Nature, with which, to read the book, he was boundlessly surrounded. His father died in 1868, and Hudson moved to London a few years later, probably in 1874. His brother had already introduced him to Darwinism, and in England he became one of the best known nature-writers of his day, although he remained destitute and bitter to the end, which came in 1922. His best work, his most passionate and pristine writing, was in *The Naturalist in La Plata* and *Idle Days in Patagonia*, but he also wrote novels. *Green Mansions*, a mystical work about a young man's pursuit of a wild child of nature, brought him a thin portion of fame, and Argentina's other most famous writer, the ever-enigmatic Jorge Luis Borges, called *The Purple Land* "perhaps unexcelled by any work of gaucho literature," which from Borges was high praise.

For Hudson, birds were a kind of salvation, as if by their flight feathers he himself was lifted out of poverty and despair. Although he confessed he could not see how such lively, swift-moving creatures could have been descended from sluggish, wrinkled reptiles, he accepted that they were, and delighted in them. To him, birds were the very embodiment of Nature. His great triumph as a naturalist was that he somehow managed to combine Wordsworth's aesthetic appreciation with the new science of Darwinism: "The rising and setting sun, the sight of a lucid blue sky after cloud and rain, the long unheard familiar call-note of some newly returned migrant, the first sight of some flower in spring, would bring back the old emotion and would be like a sudden ray of sunlight in a dark place—a momentary intense joy, to be succeeded by ineffable pain."

In *Far Away and Long Ago*, I read the strange tale of a condemned prisoner during the time of the dictator, General Rosas. The prisoner wrote a song about a bird, a ballad called *El Cuento del Bien-te-veo*, which pleased Rosas so much when he heard it "that he pardoned the condemned man and ordered his liberation." Whenever gaucho minstrels came to Hudson's house in Patagonia, he would ask them to sing it, but although everyone had heard it, no one could remember how it went. According to Hudson, the Bien-te-veo in the song was the tyrant-bird (tyrants, represented in North America by kingbirds, thus share their etymological as well as their ancestral roots with *Tyrannosaurus rex*), "a little larger than our butcher-bird," Hudson writes, "and, like it, partly rapacious in its habits."[1] This tyrannical bird seems to have taken on the role of the Trickster in Argentine folklore, rather like Coyote in Native American and Raven in Native Canadian traditions. It was always getting into trouble with the other animals and escaping from the consequences with its wit. "Old gauchos used to tell me that twenty or more years ago one often met with a reciter of ballads who could relate the whole story of the Bien-te-veo," Hudson wrote, and although he begged many old gauchos to ransack their memories, he never found one who knew the famous bird-ballad, and in the end he gave up the quest.

WE ARRIVED JUST before noon. Neuquén is about halfway down Argentina, almost exactly between Buenos Aires and Tierra del Fuego, and well inland, almost at the foothills of the Andes.

---

1 I couldn't find anything called a tyrant-bird in *The Birds of Argentina and Uruguay*, which I had also bought (for $85 US!) in Buenos Aires. There were, however, two possibilities: the great kiskadee is known locally as *benteveo común*, the common benteveo, and the three-striped flycatcher is *benteveo chico*, the small benteveo. Flycatchers, kingbirds and kiskadees are all members of the family Tyrannidae, but neither benteveo is a tyrant.

Argentina is an enormous country, a fact we Northern Hemispher-
oids often miss, accustomed as we are to seeing the world as a Mer-
cator projection, with everything north of the Equator magnified and
everything south of it diminished. Argentina is wide at the top, tapering to a point at the bottom, and if you flipped it upside down and set
it on top of North America, Tierra del Fuego would be in Labrador,
its northern border would be in Cuba, and Buenos Aires would fall
somewhere near Lafeyetteville, South Carolina. Neuquén would be
in southern Indiana—although, because it is in the foothills, a fairer
comparison might be farther west, say Winona, Kansas.

From the air, the town appeared to have been built on sand.
Winding, khaki-coloured roads looped out from the city's central
grid, and along these moved tiny, yellow dump trucks loaded with
sand. The scene reminded me of the piles of foundry sand in the
welder's yard beside our old house in Windsor, only someone had
got a set of Dinky toys for Christmas. There were sand gullies
carved in the sand, sand hills rose above the sand, and trees, all
Lombardy poplars, growing in the sand. I recalled Hudson saying
that Lombardy poplars had been planted all through Patagonia as
windbreaks because of their ability to resist strong wind. These were
resisting strong wind very vigorously as I watched them from above,
many of them bent nearly horizontal by it, the bright undersides of
their leaves flashing in the sunshine, and they sprang upright very
athletically whenever there was a pause in the onslaught. Huge
trucks were carrying sand everywhere, which I found odd, since the
wind was already doing such a fine job of it. I wondered if all the
sand around had been trucked in. Hills of sand, plateaux, buttes and
mesas of sand, sand spilling into the streets and public places, and
drifting against the curbs like tawny snow. The airport was named
Aeropuerto Juan Domingo Perón, a reminder that Perón was still
admired by many Argentines, particularly the working classes, and
he had evidently been kind to the oil industry.

I had made a reservation at the Holiday Inn, which turned out to be five hundred metres from the airport. A shuttle bus was nonetheless waiting for me. When I checked in to the hotel, I asked the clerk at the front desk to call me a cab to take me into town and went upstairs to deposit my luggage. When I returned to the lobby, the cab was waiting. As I climbed in, I noticed two large, brown-and-white birds strutting across the grass in front of the hotel, and asked the driver to wait while I looked them up in my *Birds of Argentina and Uruguay*. They were southern lapwings.

"You like birds?" said the driver, speaking English.

I told him that I had come to Patagonia to look for dinosaurs.

"Ah, dinosaurs," he said, nodding as casually as if I had said I had come to look for sand. "This is the right place. Even I found a dinosaur once. I was working for an oil company, a road construction. I drove a big machine that pushes the floor."

"A grader?" I asked.

"Yes, a grader."

"Or a bulldozer?"

"Yes, a bulldozer. I pushed over a big earth, and one of the bosses was a *geologico*, he took that big earth away and cleaned it with a brush. He took all day. At the end he said, This is a bird." He took his hands off the steering wheel and made a circle with them about the same size as the steering wheel. The cab veered sharply towards the ditch. "A chicken."

"Was it a bird or a dinosaur?"

"Then another *geologico* came from the university, and he said it was a dinosaur, not a bird. I don't know. They took it away and put it in a museum."

"Did it have feathers?" I asked.

In the mirror I could see him look at me as though he were beginning to suspect he had picked up a madman. "Feathers?" he said. "It was earth. You know, bones made of earth. No feathers."

"Rock," I said.

"Yes, that's it. Rock."

"Or fossil," I said.

"Yes. Fossil."

I sat back and thought about it. He asked me where I was going when I left Neuquén, and I told him Plaza Huincul.

"Plaza Huincul!" he said. "That's where the museum is. I'll take you there!"

"How far is it?"

"A hundred kilometres, no problem. When you want to go?"

"Not now," I said. "I'm waiting for friends. I'll let you know."

NEUQUÉN WAS A LOW town, by which I mean it clung low to the ground as though afraid it would blow away if it straightened up. It was built at the confluence of two rivers, the Rio Neuquén and the Rio Limay, where after descending from the Andes they joined to form the Rio Negro, but I never saw either river the whole time I was there. It was an oil town, unconcerned with rivers. Wind whipped down the streets, swaying the shop signs that hung out over the sidewalks like stiff laundry. The taxi driver let me off in front of a small plaza ringed with gift shops and expensive clothing stores, thinking that was what tourists liked, and told me that when I had finished browsing in there I should walk up Avenida Argentina, where the better restaurants were, then call him and he would take me back to the hotel. He gave me his card. As soon as he was out of sight I left the plaza and walked along the street proper, poking into hardware and clothing stores, paper shops and bookstalls, and peering into the lobbies of three or four small hotels, thinking that if I hadn't arranged to meet Dave Eberth and Paul Johnston at the Holiday Inn I would have moved into one of these. There were a few things I still needed to buy, some gloves, a warmer jacket, a canister of propane for my cookstove. I did turn up Avenida Argentina, crossed an open field and some railway tracks and walked to a more upscale end of town, and I did have lunch in a better restaurant. I ordered a steak and a bottle of Quilmes beer, and when they came they were both enormous. When I returned to the shops, I found them closed, according to Argentine custom; signs on the doors said they would reopen from four o'clock until nine or ten.

While I was standing in front of a hardware store, a dark-skinned woman came up to me. She had two children with her, and was pregnant with a third. She spoke to me in rapid Spanish. I told her I could not understand, *"No hablo español."* The woman simply took my hand and placed it on her belly and kept it there as she talked, as though her body was a switchboard through which her words would be transferred from her unborn child into my hand. With my other hand, I took a peso out of my pocket and gave it to her. She took it without looking at it, still talking. Then she opened my palm and began telling my fortune. I made out "You have travelled far," but little else. She lifted the hem of her skirt, and I saw that she had sewn some kind of plant into it. I said: *"Qué es?"*

"Ruda," she replied, breaking off a sprig and placing it in my hand: rue, an herb that looked like wild sage and had been used in medieval times to ward off witchcraft. With the peso I had given her she made a sign of the cross over the rue, then over me. I was becoming spooked, trapped in a doorway by this earnest woman, her wide-eyed kids, the empty street. As she spoke she patted one of my pockets—did she want more money? I told her no, I was sorry, *"Disculpe me,"* and backed away. She began speaking more rapidly, grabbed at my arm, pressing the rue into my hand, but I pulled away as gently as I could and left. She stood at the entrance to the store, looking mournfully after me and shaking her head. I suddenly had the distinct impression that it was not money she was after. I felt as though I were walking off into some awful fate that she had been trying to warn me against.

I didn't know when Paul and Dave would arrive, and when I got back to the hotel they were not there. Maybe this was the source of my disquietude, I thought. Our arrangements had been hasty and vague. Paleontologists seem to do a lot of wool-gathering, walking about as they do carrying hundreds of millions of years in their heads, imagining the comings and goings of species that came and went in a world so vastly distant and different from our own that it might as well have taken place on another planet altogether, which is why they sometimes tend to make wrong turns, miss planes and blunder into revolutions. I went up to my room and tried to read Jean Rhys's *Smile, Please*, which I had bought at a bookstore in Neuquén, one of the few English books in the store. In Buenos Aires, when I bought the bird book and the Hudson memoir, the clerk had handed me the books in a plastic bag, but here in Patagonia the art of wrapping purchases was still practised. You took a book off the shelf and handed it to a clerk, who made out a sales slip, placed the slip in the book and handed both back to you. You then took book and slip to the *caja*, the cashier, in this case an unsmiling, emaciated woman in her seventies who took the book from me as though I were handing her something distasteful, and she told you an amount. You paid her. She then made out a receipt and gave book and receipt to yet a third person, the wrapper, who made a neat bundle of your book with wrapping paper and string. Everyone in the shop except the cashier was cheerful, even to the point of pretending to understand my Spanish. The clerk held the door for me as I left, wishing me a good day. The whole transaction almost cancelled out the feeling of dread that the Black Cassandra had cast on me. Almost. What if Paul and Dave had arrived before me, and gone on to Plaza Huincul? What if they weren't coming at all?

A few pages into *Smile, Please* I fell asleep. This was no reflection on Jean Rhys, but I had had a long day. When I woke up it was dark in the room, and rather than turn on a light, I turned on the television. I always watch more television in hotel rooms than I do at home; perhaps it's the false familiarity that television engenders, that blue light making everything in the room seem ordinary. On the other hand, I was in a Holiday Inn: everything in the room *was* ordinary. When the television came on, instead of watching CNN, I switched to the Tango Channel. Twenty-four hours of tango music, with lessons and interviews with famous tango singers and dancers. All day, every day. Tango is a Buenos Aires thing, and when I was in Buenos Aires I had avoided it as assiduously as one avoids the can-can in Paris. Most of the famous tango bars, the Hippodrome, for example, had been turned into discotheques anyway. But now that I was safely in Patagonia it was almost exotic, and I could pretend I was improving my Spanish at the same time. According to Borges, tango was invented in the brothels of Buenos Aires sometime in the 1880s. Like New Orleans jazz, it was the authentic, animalistic cry of the streets, the literature of the illiterate, the secret language of those who had no voice. It was about sex and violence, submission and aggression. Wealthy *porteños*, as the middle-class citizens of Buenos Aires are called, ignored it until it was taken up in Paris, after which they reclaimed it in its more sanitized form as the true music of Argentina.

In the brothels it had been played on pianos, flutes and violins. The guitar and bandeneons (small accordions with buttons instead of keys) came later. The orchestra on the television had two violins, two bandeneons, a piano and two guitars. The singer was a slick-haired man named Hector Pacheco, whose eyelids drooped as though they were a Darwinian adaptation to cigarette smoke. The dancers moved with exaggerated control, swirling so quickly and stiffly that they had to cling to each other to counter the centrifugal forces trying to fling them apart. Their clinging was what the *porteños* had disliked so much at the beginning; they called the

woman the bottle and the man, the cork. To me, they were like figure skaters without skates. There were illustrations breaking the
dance down into its component steps: *la marca, la pompa, la cami-
nata con giro* and *el bol masculino*. Eduardo Aguirre demonstrated
the *salida con arrepentida*, which according to my pocket dictionary
meant "exit with repentance." Aguirre was a squat, balding man
with a low centre of gravity, shaped something like a bowling pin.
His female partner was at least a head taller than him, and he
clasped her in an iron grip so that she was like a caged bird, and
the dance took on the aspect of a long, choreographed attempt at
confinement and escape. From the knees up she was as rigid as a
department-store mannequin; all her rebelliousness was concentrated in her feet, which seemed to have wills of their own, for they
jerked and spun and flew up at rakish angles as the couple danced,
as though she were surreptitiously trying to kick Aguirre in the
crotch. I could easily imagine her stabbing him with a silver *estilete*
without a second thought. No wonder he watched her so carefully.
I watched her myself for about half an hour, just to see if she would
break loose, but she never did, and eventually I executed a *salida con
arrepentida* of my own by falling back to sleep.

IN THE MORNING, after my free continental breakfast in the hotel
lobby beside the giant-screen TV (*café con fútbol*), I went for a walk
with my binoculars and bird book. Between the hotel and the airport,
a narrow strip of land incorporated a gravel road, a railway line and
an irrigation ditch bordered by parallel rows of poplars. I took a path
leading down to the ditch where, almost immediately, a plumbeous
rail swam out from under a rusted iron bridge, looking for all the
world like a demented chicken. In a field beside the path there were
more southern lapwings, which seemed to fill the niche occupied at
home by crows and magpies. There was also a great kiskadee and, in
the distance, something that looked like a hawk perched on a clump

of sand-coloured earth. I decided it was probably a hawk-shaped clump of earth. It was too hot to check. I thought that if I walked far enough I would reach Neuquén, but after half an hour I found myself at a crossroads standing outside a small store, someone's living room, really, from which newspapers, lottery tickets and fish were sold. I bought the local paper, the *Rio Negro*. I was obviously still a long way from the centre of anything. The sun was already high and I was parched and dusty, and there were no more birds. I hailed a taxi that was parked by the curb and told the driver to take me into town.

In Neuquén I went into a restaurant, ordered a *café con leche*, and sat at a table near a window with my little dictionary and the newspaper. Why not have another go, I thought. Apple farmers from three districts north of the city appeared to be demanding subsidies of three thousand pesos per hectare for the loss of crops destroyed the week before by little pebbles falling from the sky, or maybe hailstones; they had driven their tractors into Neuquén and were blocking the roads around City Hall until their demands were met. At the Cinema Reál on Avenída Argentina I could see Mel Gibson in *La Ravanche* or Robin Williams in *Patch Adams*. Halfway through the paper, a headline caught my eye: "*Rescatan un 'dino' carnívoro de 80 millones de años*," which I took to mean that an 80-million-year-old carnivorous dinosaur had been rescued from somewhere. Laboriously, I worked my way through the first paragraph. Spanish is really a simple language when you sort of unfocus your mind to it. "*La cola de un dinosaurio carnívoro afloró en la superficie del Auca Mahuida, y así pudo ser rescatado del sitio en el descansó durante cerca de 80 millones de años.*" The tail of a carnivorous dinosaur had risen to the surface at Auca Mahuida, whatever that was, and had been recovered from the site where it had lain hidden for the past eighty million years. I sailed on: a team of *paleontólogos* and *técnicos* had apparently been working in the area since last year, when they "revolutionized the international scientific world" with their discovery of the largest dinosaur nest known to science, and three weeks later found a complete skeleton of something, maybe an embryo. Whatever it was they found this year was now relaxing in the Carmen Funes Museum in Plaza Huincul, wearing a plaster jacket. Plaza Huincul was two hundred kilometres from Auca Mahuida.

I skipped down a few paragraphs to a description of this year's find. It was an almost complete skeleton of an animal closely resembling *Carnotaurus*, one of the South American theropods that resembled North America's tyrannosaurs. According to Rodolfo Coria, who was quoted in the article, "this specimen will provide new data and information to those who study it." Nice quote, I thought. Coria sounded like a cop in a detective novel, especially as he went on to note that the dinosaur "might have died from natural causes." The rest of the article described the area as it was thought to have looked during the Late Cretaceous, when the *Carnotaurus* lived, probably by predating upon whatever species of sauropod laid all those eggs. There were no Andes mountains then, and therefore no steppes; the area that is now Plaza Huincul was a low, flat plain cut through by large rivers that often flooded their banks as they emptied into the Atlantic Ocean. It sounded very much like Cretaceous Alberta. Water and silt from these repeated floodings may have drowned the eggs, some with embryonic sauropods still inside them, and also buried the skeletons of the carnivorous dinosaurs that came to feed on them.

I knew something about *Carnotaurus*. There had been that ridged skull in the Tyrrell's dinosaur hall, with a note saying the first one had been found in 1985 by the Argentine paleontologist José Bonaparte in a part of southern Argentina that had been dated as Early Cretaceous. This article suggested that that dating was now being questioned. Coria and Luis Chiappe, both former students of Bonaparte's, had found this new specimen in a *Late* Cretaceous deposit in *northern* Patagonia. If it turned out to be the same species as Bonaparte's, it would mean that *Carnotaurus* had lived nearly fifty million years more recently than Bonaparte thought it had.

There were several other remarkable features about *Carnotaurus*. It was small for a large theropod: twenty-five feet long and weighing about a ton, with a bull-like neck and a small head. Its skull ridges were actually horns protruding over each of its eyes. Such horns are unknown in any other species of theropod. Even nowadays they are found only on prey species, goats and deer and cows, animals that don't have nice, pointed teeth to defend themselves with. The horned dinosaurs, the ceratopsians and stegosaurs, were cud-chewing, myopic, reptilian versions of rhinos or bison. Why would a theropod, which had teeth that could saw through five-inch bones, need horns? Bonaparte's specimen was also remarkable because it had extensive skin impressions preserved along one entire side of the skeleton. Paleontologists don't often get skin or for that matter any other soft body parts. *Carnotaurus sastrei* had small tubercles all over its body, like giant goose bumps, surrounding low, cone-shaped warts arranged in rows along its tail, up its flanks and back up to the head. Could they have supported feathers, I wondered? Perhaps this new find would provide more details, and I was looking forward to asking Chief Inspector Coria about it when I got to Plaza Huincul.

THINKING ABOUT PLAZA HUINCUL made me anxious to get there, so I ordered another coffee and sat in the restaurant trying to figure out what I would do if Dave and Paul didn't show up soon, or at all. Science, like travel, is the process of moving from the known to the unknown, and Neuquén was definitely becoming known. It was a nice enough town, but I was impatient to get on with the journey. I began to wonder what was keeping the others, and how we were supposed to hook up with Phil and the rest of the expedition members. I hadn't heard anything for weeks before I left home, which, now that I thought about it, was in itself worrisome. No news is not good news. It was possible that there had been a change in plans, that the dig was postponed for another month, or cancelled altogether, and no one could reach me to tell me about it. Other things come up, funding gets cut, kids get sick. I wondered what I would do if, after a few more days, no one appeared and no messages arrived for me. Would I make my way back to Buenos Aires and go home? No, I thought, I'd go on to Plaza Huincul on my own, introduce myself to Rodolfo, and offer to work with him for a few weeks. Cleaning bones, shovelling gravel, anything. With this pleasant prospect in mind, I left the restaurant and walked up to the bus station, which I found to be a long, flat, dusty building squatting in a field of sand between the old section of town and the tonier area along Avenida Argentina. A row of dented and greasy buses were nosed up to its belly like piglets at a prostrate sow. Most of the piglets bore the logo of the local bus company, El Petroleo, but a few were tour coaches, double-decked kneelers, painted metallic mauve or chrome yellow, with curtained windows. Exotic destinations scrolled across black screens above their windshields: Bariloche, Bahia Blanca, San Martín de los Andes. Inside the station were handwritten schedules above a line of ticket windows, and I could see without asking that there were several buses a day to Plaza Huincul, and the fare was only five and a half pesos. There was a lunch counter in the waiting room that sold *empanadas* and cold beer. This seemed a positive omen. I realized I was already looking forward to making the rest of the trip on my own. This was Friday. If there was no word from Paul or Phil by the end of the next day, I decided, I'd leave for Plaza Huincul on the noon bus on Sunday, get a hotel in Plaza Huincul, if there was one, and call Rodolfo first thing Monday morning. Having a plan, even a hare-brained one, made me feel less anxious. As I left the station, I noticed that one of the smaller buses belonged to the Holiday Inn. I went over to it and recognized the driver from the airport two days ago who had driven me the five hundred metres to the hotel; he said there was a free shuttle service between the hotel and Neuquén. Things were definitely looking up, I thought, as I climbed aboard.

WHEN PAUL AND DAVE didn't arrive on the 12:30 flight from Buenos Aires on Saturday, I settled in the hotel lobby with W. H. Hudson to await the 4:30. The corner of the lobby was a bar, and the bartender, after taking my order for my first beer, told me that if I wanted another I should just help myself to the cooler on the other side of the room and tell him how many I'd had before leaving. This seemed an excellent disposition of things, especially as it did not seem that it would bother the barman unduly if I lost count, which I determined not to do. Four-thirty came and went. The next and last flight was at 8:30. I plodded on through *Far Away and Long Ago*.

For Hudson, Patagonia remained "an enchanted realm, a nature at once natural and supernatural." Science and religion, Darwin and Wordsworth. But the book started with an enigma and ended with an anachronism, and I found myself not trusting it. Although his biographers say Hudson was born in Quilmes and spent only a short time in Patagonia, Hudson himself says he was born "on the South American pampas" in a house that stood on a high elevation, surrounded by limitless horizons and waving grass and the famous *ombú* trees that have come to be associated with his name. This didn't sound like Quilmes to me. And in the the last chapter, he writes that he read Darwin's *The Origin of Species* when he was sixteen, even though he was born in 1841 and the *Origin* was not published until 1859. Throughout the book, in fact, time and place are fuzzied and convoluted, woven in and out, spiralling forward and leap-frogging backward. Hudson wrote it propped up in bed with a high fever during six weeks of long, sleepless nights, while his nurse dozed in a corner by the fire, and so we should not place too much faith in the accuracy of his memory. He himself acknowledges that memory is selective and elastic, barely removed from fiction: "The scenes, people, events we are able by an effort to call up do not present themselves in order; there is no order, no sequence or regular progression—nothing, in fact, but isolated spots or patches, brightly illumined and vividly seen, in the midst of a wide, shrouded, mental landscape."

He recalls his reaction to reading Darwin, however, with particular clarity. The copy of the *Origin* was lent to him by his elder brother, who had just returned from England and wanted to know how Hudson could continue to believe in God in the wake of Darwin's theory of evolution. Hudson read the book and replied that "it had not hurt me in the least, since Darwin had to my mind only succeeded in disproving his own theory with his argument from artificial selection. He himself confessed that no new species had ever been produced in that way." This was the same doorstop over which Huxley had stumbled; was it possible that in his feverish state Hudson confused his own reading of Darwin with Huxley's? His brother replied that Hudson's observation "was the easy criticism that anyone who came to the reading in a hostile spirit would make," and that Darwin had answered it in his great work. "Read it again," he told the young Hudson, "in the right way for you to read it—as a naturalist." And so Hudson reread Darwin, and then looked again at the birds and mammals about him in Patagonia, the same Patagonia that had inspired Darwin's theories in the first place, and became an ardent evolutionist. Each individual he encountered, he suddenly realized, "was a type, representing a group marked by a family likeness not only in figure and colouring and language, but in mind as well, in habits and the most trivial traits and tricks of gesture and so on; the entire group in its turn related to another group, and to others, still further and further away, the likeness growing less and less. What explanation was possible but that of community of descent?"

The more I read the more convinced I became that Hudson had faked his own autobiography, that he was making up this enthusiastic response to Darwin in his boarding house in England. He seemed particularly anxious to portray himself as an early convert, and I thought I knew why. In later editions of the *Origin*, which Darwin was continually expanding and amending, he referred to Hudson as "an excellent observer" but "a strong disbeliever in evolution." This remark must have stemmed from his reading of a series of letters published by Hudson in 1870 in the *Proceedings of the Zoological Society of London*, to which Hudson, then still living in Argentina, was a young field correspondent. These letters were entirely about Argentine birds, and later formed the basis for his massive *Birds of La Plata*. In Letter 2, however, he alluded to a bird called the "pampa woodpecker," which he said was a Carpintero, the Spanish word for woodpecker. He added that he had often seen this bird perching on trees, despite the fact that Darwin, in the *Origin*, "has so unfortunately said—'It is a woodpecker which never climbs a tree'!" Not content with this sideswipe, he renewed the attack in Letter 3: "However close an observer that naturalist may be," he wrote, meaning Darwin, "it was not possible for him to know much of a species from seeing perhaps one or two individuals in the course of a rapid ride across the pampas. Certainly, if he had truly known the habits of the bird, he would not have attempted to adduce from it an argument in favour of his theory of the *Origin of Species*, as so great a deviation from the truth in this instance might give the opponents of his book a reason for considering other statements in it erroneous or exaggerated." Well into his stride now, Hudson went on to virtually accuse Darwin of deliberately falsifying data in order to support his hypothesis. To one familiar with the habits of the pampa woodpecker, he continued, "it might seem that [Darwin] purposely wrested the truths of Nature to prove his theory; but as his 'Researches,' written before the theory of Natural Selection was conceived—abounds in similar misstatements, when treating of this country, it should rather, I think, be attributed to carelessness."

Darwin's replies to these charges, printed in later editions of the *Proceedings*, were gracious. He had, he wrote, in fact spent many months observing thousands of *Colaptes campestre* (giving the bird its correct name; it was a flicker, not a woodpecker), had based his remarks on painstaking research, and moreover no less an authority than the renowned Paraguayan ornithologist Luis Azara confirmed that "it never visits woods, or climbs up trees, or searches for insects under the bark." Darwin concluded, more in sorrow than in anger: "I should be loath to think that there are many naturalists who, without any evidence, would accuse a fellow worker of telling a deliberate falsehood to prove his theory," which is exactly what Hudson had done.

Hudson never allowed these letters to be reprinted in subsequent collections of his writing. Perhaps he found them embarrassing. We have all written things in our youth that we would rather not be reminded of, and certainly not have reprinted, not because we were wrong, but because it is no longer important that we were right. But their existence, and Darwin's cutting remark in the 1871 edition of the *Origin*, must have made him wince, and it is not surprising that he would want to have the final word when composing his memoir.

As I finished Hudson's sad and saddening book, I looked up and became aware of a familiar figure standing at the reception desk: Paul Johnston, wearing a floppy hat and a blue nylon jacket and surrounded by backpacks and kit bags and a group of other passengers from the shuttle bus. Dave Eberth wasn't among them, but now that Paul had arrived I breathed a sigh of relief. The expedition was on.

CHAPTER FOUR

Theropod Heaven

IN DRUMHELLER, PAUL had seemed to me to trail around an aura of unnatural calm, as though inside he was excited about something the rest of us couldn't see. He was thinking, he'd told me then, about two big projects. One involved disproving a long-held theory that certain perforations in fossil ammonite shells had been made by marauding mosasaurs. Ammonites lived in those curled shells that look like fossilized ram's horns and keep turning up in farmers' fields—polished and attached to neck chains they are sold as "the Alberta gemstone." Mosasaurs were huge, crocodile-like predators whose needle-sharp teeth certainly could have crunched through the soft shells of living ammonites, but Paul didn't think they had, and to test his theory he had built a large, mechanical mosasaur jaw out of wood and pneumatic pumps, and crushed modern shells in it to see if it made the same kind of holes. It didn't. He concluded that the holes had been made by small, burrowing invertebrates known as limpets. Now the whole world of invertebrate paleontology was lined up on either side of the mosasaur-limpet line, and Paul was the leading apostle for the limpet faction, a role he seemed to enjoy in his quiet, intense way. His other project was still in the hunch stage: he suspected that life on Earth had initially begun, not in masses of blue-green algae floating on a sun-swept sea, as has been maintained, but rather in association with certain deep-sea vents, fissures on the

75

floor of a vast, oxygenless, frozen ocean, where gases and various minerals seeped from the planet's molten core. He thought the weird ancestors of the Burgess Shale fauna might have hovered near these vents, breathing some other gas than oxygen through organs that were not gills. These were definitely other-worldly thoughts to be carrying around, and Paul seemed happily detached from the here and now as he mulled them over. Switching from one to the other must have been a bit like taking a break from trying to solve Fermat's Enigma to play a chess match with Bobby Fischer.

He'd only recently learned he was coming to Patagonia. "Phil's way," he'd said, "is to poke his head in your door and say, 'Do you want to go to Argentina?' and I say, 'Yeah,' and he says, 'Okay, you're on,' and I say, 'What for?' He hasn't answered that last question yet."

And here he was anyway. When he finished registering at the desk, he looked around the room, made his way over to my table and sank wearily into a plastic chair. He'd been travelling for twenty-two hours, from Calgary to Vancouver, through Los Angeles to Buenos Aires, and from there to Neuquén, and his body was surprising him by being tired.

"Is Dave coming later?" I asked.

"Dave's not coming at all," said Paul. "It's a long story."

Before I could inquire further we were joined by a man who had been making his way across the lobby by kicking his luggage ahead of him along the floor. He was in his late twenties, I guessed, wearing faded blue jeans, a blue Polar Fleece sweater zippered up to his throat, and huge leather hiking boots. I thought of Paul Bunyan, especially as there was a large, wooden-handled pick strapped to the bulkiest of his three packs. His head was shaved almost smooth, he wore wire-framed glasses and had a small ring in his left ear. Maybe not Paul Bunyan; maybe a professional wrestler. But his eyes were deep and glowed with intelligence and energy. "This is J.-P.," Paul said. "He's here instead of Dave."

"J.-P.," I said, shaking his hand. "Short for John-Paul?"

"Just J.-P.," he said, sitting down. The waiter came from behind the bar and over to our table, and J.-P. looked up at him. "Do I want a beer?" he asked himself. "It'll probably knock me out. What time is it? Holy shit, it's still early. Okay, I want a beer. *Cerveza, por favor*. And some food. Do they serve food here?" He looked at the menu, which was typed in both Spanish and English, and consisted mainly of sandwiches. In Argentina, sandwiches are cut from large squares of bread, toasted and pressed so flat they looked as though they've been clamped in a bookbinder's press, and always involve cheese and fried eggs. Paul and J.-P. both ordered beef sandwiches.

"Dave took a job at the museum in administration," Paul said. "He's gone. No more geology, no more field work. Amazing, isn't it? And it wasn't just thrust on him, either; he *applied* for it." We all shook our heads, although for Paul and me it was more in sadness than in disbelief. His decision seemed, in retrospect, to be completely predictable. I remembered my last conversation with him in January in his office, when he'd reflected on the impossibility of a field paleontologist having any kind of normal home life.

J.-P. had learned he was coming to Patagonia only three weeks before. He was a post-doctoral geology student at the University of Calgary who taught a bit, did a bit of work for the oil companies, and occasionally took geologists on field trips into the Rockies, where he'd become interested in fossil lobsters. When Phil called him and asked him if he wanted to do some geologizing in Patagonia, he'd sighed heavily, consulted with his wife, Shima, to whom he'd been married a little more than a year, and said yes. "Then I realized I knew dick-all about Patagonia, except that it's probably non-marine. At least I think it's non-marine. Maybe it's marine. I'm not going to do much speculating about it now, though; just measure the shit out of it and think about it when I get home."

"It might be non-marine with a marine base," Paul said speculatively.

The rest of their conversation took the form of a lot more spec-ulation, it seemed to me, larded with the kind of words only geol-ogists understand, words like parasuccession and transtemporal. Paul and J.-P. seemed an oddly matched team, Paul older and calmer, lounging in his chair, letting his mind play over possibili-ties without taking any of them seriously until the evidence was in, J.-P. sitting up straight, crushing beer cans in his immense hands, tense and argumentative, throwing out ideas like chal-lenges. I wondered what their conversation had been like for the past twenty-two hours. Then I felt my mind blurr off into a Quilmes-induced reverie. I imagined myself speaking fluent Span-ish, sitting with Borges in his darkened apartment, in the perfect twilight of his blindness, three of us, me and Borges and Paul Theroux, discussing aspects of Patagonia. Maybe Bruce Chatwin was there, too, telling us that the name Patagonia was taken from a sixteenth-century Spanish epic poem in which the beast Pata-gon is hunted down in a far-off land by a Spanish knight; the poem and the beast inspired Shakespeare's *The Tempest*. The land of Patagon has been a semi-mythological state since it was first visited by Ferdinand Magellan in 1520. Elizabethan mariners knew it as a forbidding realm of black fogs, roiling waters, blind-ing snowstorms and numbing cold, inhabited by shaggy giants they still called Patagons, and even stranger animals: flightless birds, armoured rodents, swimming cows.

"There is nothing in Patagonia," Borges told Theroux, but he was wrong, or he may have said it as a diversion, as a parent might tell a child, "There's nothing in that closet." In 1577, Francis Drake, stopping for water on his voyage around the world, ventured inland a few leagues and found himself in an eerily empty land; then he saw "the footing of people in the clay-ground, showing that they were men of great stature," and hastily retreated to his ship. The footprints were those of the Tehuelche, the Patagons, a giant people reputedly without the power of reason or speech or, apparently, the ability to feel cold: men and women paddled naked in their canoes through the icy waters. Ten years later Thomas Cavendish, following Drake's route almost precisely, encountered the breeding places of seals with manes like lions and feet like a man's hand; when shot with arquebuses, the seals sustained no visible injury. On the mainland they saw savages "as wild as ever was a buck or any other wild beast." Cavendish's crew shot several of them and measured their feet; they were eighteen inches long. Mammoth men. Everything in South America is mammoth, the biggest in the world. A Canadian Baptist missionary, George Whitfield Ray, came to Argentina in 1890, an experience he recorded in his book *Through Five Republics on Horseback*, which contains a black-and-white photograph of a hacienda-sized boulder poised precariously on the brow of a granite outcrop, as though about to plunge onto the plains far below; the caption reads: "The World's Largest Rocking Stone, Tandil, Argentina." Another shows a lake covered with flat leaves, their edges turned up and fluted, like serving trays: "Victoria Regia, The World's Largest Flower." Yet another is of a huge waterfall, reminiscent of Niagara but captioned "The Falls of Yguasu, The Largest Falls of the World." Patagonia was the southern hemisphere's Ultima Thule, the biggest, farthest, most desolate, most unencompassable place in the world.

Even Darwin, who found something interesting pretty well wherever he looked, was put off by Patagonia's implacable vastness. "The surface is everywhere covered up by a thick bed of gravel, which extends far and wide over the open plain. The vegetation is scanty; and although there are bushes of many kinds, all are armed with formidable thorns, which seem to warn the stranger not to enter on these inhospitable regions." He was writing about the flatlands around the mouth of the Rio Negro. And here we were, at its source, one thousand miles upriver.

PHIL AND THE OTHER CANADIANS were due to arrive the following day at two in the afternoon, on a flight from Santiago. That gave us the morning to do our last-minute shopping in Neuquén. We took the 8:30 shuttle into town. Paul wanted a haircut and a bottle of 7-Up; J.-P. wanted cookies and a bottle of Glenfiddich; I still wanted a canister or two of propane, a small kettle, some tea and a pair of gloves. The temperature was dropping daily, for we were in the foothills and it was the Patagonian autumn, a time of cruel winds and cold rain. The papers were already reporting snow in Bariloche, Patagonia's Banff, only an inch or so south of us on the map. We left Paul at a beauty salon across from the bus station, in a crowded plaza of open-fronted shops selling everything from shoes and combs to newspapers and hot dogs.

"How do I tell them how I want my hair if I don't speak Spanish?" he asked nervously.

"What the hell," J.-P. said, pushing him through the door, "we'll be in camp for a month: whatever they do to you will grow out by then."

There was a clothing store next to the salon, where I bought a winter jacket. I tried asking the clerk if she sold gloves, *guantes*, to which she replied, I think: "No, they don't come in until after Lent." J.-P. was in a souvenir shop next door, peering into a display case full of rocks and minerals. "Quartz, mostly," he said, disappointed. "Definitely no bone." We went into the hardware store in front of which, two days before, I had been button-holed by the woman with the rue in her hem, but she was nowhere in sight. I bought a canister of propane, and when we turned a corner we found ourselves outside a gigantic supermarket called Tía. We went in and found everything else we needed (except gloves), and when we came out again Paul was standing on the corner, running his fingers through his hair. "How does it look?" he asked. It looked unchanged to me, but J.-P. said: "Like a shampoo commercial," and Paul looked pleased. "That's how I told them what I wanted," he said. "I pointed to a shampoo ad in a magazine."

I looked at my watch. The morning dampness was still on the ground, and although sunlight was slanting bravely in from the north, there was an autumn chill in the air. I put on my new coat and thought again about gloves, wondering what they had to do with Lent. "Maybe we should have lunch early," I said. "We have to get back to the hotel, check out and be at the airport by two." We found a restaurant on Avenida Argentina that was open, had another meal of flattened egg-and-beef sandwiches, and took a taxi back to the Holiday Inn. By 1:30 we were sitting in the arrivals lounge at the airport, barricaded by luggage and trying to discern the word "Santiago" from the sputtering noise occasionally emitted by the airport's public-address system. The lounge was crowded with people dressed in their best clothes, kissing and hugging each other goodbye or hello. Everyone seemed happy to be on the move, I thought, and I was even happy to be associated with movement, to be in an airport but not waiting to get on a plane. People generally behave well in airports, kissing each other hello or goodbye, mindful of their children and their suitcases. We took turns sitting with our bags, but none of us knew what Rodolfo looked like so there wasn't much point in wandering through the crowd looking for him. I went up some stairs leading to a mezzanine level, where there were souvenir shops and newspaper kiosks, but was not inclined to buy anything. The side of the building facing the runways was all windows, some of which were open, and I watched a few planes fly in, landing as gracefully as geese on a still pond, and then taxi towards us like awkward iguanas on a pebbled beach. Among the passengers alighting from one of them I recognized

Phil's tall, blond, gangly figure, and hurried downstairs to join the others. The next leg of our journey was about to begin.

WHEN I REACHED THE arrivals area, Phil and Eva were already through security. When Eva saw me, she came over and gave me a brief hug. We had met in Drumheller and she had invited me to their house for dinner. She is a palynologist, a paleontologist who specializes in fossil pollen, and is originally from Denmark. She and Phil met in 1993 when he was in Denmark to deliver a paper on Chinese dinosaurs at the 150th anniversary of the Danish Geological Museum. Eva was working for the Danish Geological Survey at that time; she'd gone up to him after his paper and asked him what China had been like, and when he got back to Drumheller he sent her a copy of my book, *The Dinosaur Project*. She read it that summer in Greenland, camped on the side of a bald, frozen mountain, collecting rock samples by day, crushing them in her lab at night and checking them through a microscope for fossil pollen. When she was back in Copenhagen, she sent Phil a postcard telling him she'd like to see him again if he ever came back to Denmark, and shortly after that Phil bought an airline ticket. "It was your book that brought us together," Phil told me at dinner.

When his divorce came through, Eva came to work at the Tyrrell Museum, and they'd been working and living together ever since. Eva's son Rasmus comes to stay with them in the summers. He and Phil get along well, because both of them are interested in music. Erasmus is studying classical piano; Phil's taste tends towards heavy metal, but he has a fine collection of classical and blues. There is never not a CD on the player at home: Phil keeps a computerized record of every CD he listens to, and charts it against what else is happening in his life, to see if there is a correlation. There often is.

Eva took me over to a small group that had formed around Phil. Paul and J.-P. were standing with Mike Getty, Phil's field assistant from the Tyrrell, who was speaking fluent Spanish with two of Rodolfo's technicians from the museum in Plaza Huincul, Christian Giminez and Daniel Hernandez. Eva introduced me to Rodolfo, who didn't look at all like a Chief Inspector, but rather reminded me of a hipster from the 1950s. Everyone looked a bit rumpled from their various journeyings except Rodolfo, who was casually dressed but not a hair out of place: sunglasses, white T-shirt under a light windbreaker with the collar turned up, neatly ironed blue jeans. I imagined a rat-tailed comb sticking out of his back pocket, but of course there wasn't one. Only his scuffed hiking boots gave him away as a scientist. His thick, black hair was carefully trained to fall naturally over his forehead, and his quick smile revealed a lustrous row of perfect teeth. When he had acknowledged Eva's introduction, he went back to what he had been saying to Phil. There seemed to be some difficulty with a box of materials that Mike had sent from Alberta and which Rodolfo had been unable to pry out of customs. "It was addressed to me," he explained, "so when I went to pick it up, they asked me what was in it, and I said I didn't know."

"Didn't you get Mike's list?" Eva asked him.

"I didn't receive any list," Rodolfo said. "And because the box was addressed to me, and I am a citizen of Argentina, I will have to pay duty on its contents. So if I don't know its contents, they can't calculate the duty. Now we have to wait and come back on Monday, and I'm afraid it's going to cost a lot of money to get it away from those thieves in customs."

"Why's that?" asked Phil. "How much is the duty?"

"Half the value of the contents," Rodolfo said angrily.

There was a glum pause about the circle as we contemplated the alternatives. Phil turned to Mike. "How much do you figure the stuff is worth?" he said.

Mike shrugged. "I don't know. It has all our tools in it, a couple of extra tents, work gloves, all the rolls of Gypsona and burlap for making jackets. Maybe twelve hundred dollars." Rodolfo shook his head. "It's too bad," said Mike, "the whole idea of shipping the stuff down here was that it was cheaper in Canada. This is going to make it twice as expensive." Eva looked worried, Rodolfo looked embarrassed, and I was reminded that this was not an over-funded expedition.

In the end, Phil decided there was nothing to be done about the box until Monday, and we all clambered into the two vehicles Rodolfo had brought from Plaza Huincul. One was a second-hand ambulance that the museum had inherited from the local hospital, and I ended up riding in the back with Mike, the two Argentine technicians and a mound of gear piled where the patient would have been. While the three conversed in Spanish, I gazed out the window at the passing countryside. We were following a long row of Lombardy poplars, heading, I deduced, north and west, north because the sun was ahead of us, west because we were gradually climbing. We were also apparently driving into rain. Behind the poplars I sensed rather than saw a river, probably the Rio Limay. Through the line of trees I could see men with their backs to the road holding fly rods aloft, and unless they were casting over dry sand, there was a river there. Besides, something had to be nourishing the trees, and it certainly didn't appear to be the soil. Patagonia's countryside gave new meaning to the term earth tones: words like yellow, brown and tan didn't come close to conveying the richness and depth of the desert palette. Here colour took on the quality of prose, each tone containing its own history of colour, the process of producing itself: burnt sienna, red oxide, yellow ochre, ash grey. Colour formed by fire, colour born in the absence of water. Reds the texture of dried blood, yellows like egg yolk or powdered mustard, charcoal greys and baked blacks. Gazing out at the surging landscape, I understood that "tone" is also a musical term. Here and there were splashes of green, now rendered deeper by the rain. We were passing through tufts of thin grasses, low spindly bushes. I watched for birds, but we were moving too fast. Behind us, the plumb-line of Lombardy poplars plunged over the horizon.

Mike turned to me and said he'd been trying to figure out some Argentine expressions. "For example," he said, "they often put the word *che* at the end of a sentence. *Esta salsa es muy piquante, che*! Something like, This sauce is really hot, man! I asked them why *che*, and they said it was in homage to Che Guevara. Not many people realize he was born in Argentina, but he's a national hero here."[1]

As it happened, I had been reading Guevara's autobiography before I left Canada. I'd wanted to find out if he'd ever travelled in this part of Patagonia. In fact he had, in January 1951, to be precise, when he was still a young medical student named Ernesto, a middle-class *porteño* in the making. He and a friend, a fellow student named Alberto Grenado, had taken a year off university and driven along this same stretch of road on their motorcycle, an old Norton 500 which they called La Poderosa, the Powerful One. Their intention was to cross the Andes into Chile somewhere down around Bariloche, and from there make their way up the coast, through Central America and into the United States. As they were driving in the dark along this part of what was then a gravel road from Neuquén, the motorcycle broke down and the two adventurers were forced to drag it and their gear off the road and set up their tent in the very hills we were passing through now, for the night. Although it was January, high summer in these latitudes, the weather turned cold and windy, the evening's breeze stirring up into

---

1 Ending a sentence with the emphatic tag *che* was an old custom in Argentina, going back much farther than the Cuban Revolution. It is far more likely that Ernesto Guevara was nicknamed Che because, unlike his fellow revolutionaries, who were mostly Cubans, he spoke like an Argentine, ending his sentences with "che," as in, "Hey, Fidel, pass the bread down here, che," much as Canadians are supposed to end their sentences with "eh?"

a gale that ripped their tent off its pegs, exposing the two exhausted travellers to an icy, biting wind and threatening to blow the motor-cycle over a cliff. In the darkness, they tied the machine to a tele-phone pole and wrapped the shredded tent around it to keep the engine dry, and spent the rest of the night in its lee, shivering in sodden blankets.

The next day they wired the motorcycle together and pushed it twenty kilometres to the nearest town, where, while it was being welded, they stayed with a group of farmhands on the *estancia* of a German landowner, climbing fruit trees and gorging themselves on ripe plums, catching rainbow trout in a river that ran through the property, playing the part of poor students and generally alienating themselves from the sympathy of the farmhands, who found them lazy, irresponsible and hopelessly bourgeois—two supercilious "doc-tors" from the capital out on a spree. Guevara records these reac-tions without comment, no doubt secure in the knowledge that the world would find them amusing in the light of later events (he worked his diaries into a narrative much later, in Cuba, where, edited by his Cuban widow, they were eventually published by the Che Guevara Latin America Centre). The experience was a turn-ing point in Guevara's life, not, as his father would later say, because it was fraught from the beginning with "the mystical and certain knowledge of his own destiny," but because it was the first time he had spent any time with the working poor. His father was a con-struction engineer in Buenos Aires when Ernesto was born in 1928; the Guevaras were staunchly middle class and determined that their oldest son would be a solid citizen of the new Argentina. Instead, Ernesto turned out to be an inveterate wanderer. "I now know," he wrote just after his stay at the *estancia*, "by a fatalistic coincidence with fact, that I am destined to travel." In 1950 he had toured most of Patagonia on a moped, alone, and the next year set out on this trip with Grenado. This was not his road to Damascus, however; that would be the next trip, 1953, when he witnessed worker revolts in Bolivia after that country's National Revolution. The year after that, in Mexico, he would meet Fidel Castro, who was planning his invasion of Cuba, and join Castro's little army as a physician. For now, he and Grenado were on a lark. In San Martín de los Andes, the two of them liked the soft life so much they contemplated opening a clinic there after they graduated, treating tourists for ski injuries and working on their tans in the summer. "An expedition has two points," he wrote outside Neuquén, "the point of departure and the point of arrival." The part in between, the actual expedition, was of less interest. There is little description of the landscape or of the people who lived on it. For Guevara, the point of arrival was Caracas, Venezuela, where he realized he was completely at home among the city's dockworkers and stablehands. The diary ends with his declaration that he would henceforth lend his voice to "the bestial howl of the victorious proletariat."[2]

Every so often we passed tiny structures set up along the roadside, miniature, open-fronted houses made of adobe bricks. They were evidently shrines to family members who had been killed in road accidents. Some were quite elaborate, with little steps leading up to them, and glass windows set in the sides, chimneys mounted on tile roofs. Inside, someone had placed articles no doubt once favoured by the deceased: stuffed animals, plastic toys, bottles of Coca-Cola, paper flowers. Many were distinguished by small signs, some blunt and accepting (*Difunto*—Dead), others more imploring (*Nuestra Señora del Correo*—Our Lady of the Mail?). I wondered how long these simple shrines were maintained, whether grieving parents came out weekly to dust off the flowers and replenish the Coke, and if so whether it went on for a year, or a lifetime. In

---

2 The diary was published in Italian in 1993 as *Latinoamericana: Un diario per un viaggio in motocicletta*, and in English in 1995 as *The Motorcycle Diaries: A Journey Around South America*.

Buenos Aires I had visited the cemetery attached to the Claustros de Pilar, the seventeenth-century cloisters of the Recolets, who gave their name to to the city's wealthiest district. I'd entered the cemetery through huge portals and found myself in a vast, concrete necropolis, rows of houses for the dead laid out in streets and plazas, the coffins plainly visible through wrought-iron' gates or glass doors, set on benches and shelves in buildings much grander than these tiny roadside shrines, but emanating from the same desire to keep the dead among us, the paleontological impulse. The monuments in Recoleta were meant to impress the living rather than the dead, it seemed to me, for why else would the family name be emblazoned across the front of each house? Do the dead forget their names? A funeral had been in progress just inside the gates, in the rain, men in grey tuxedos and women in long gowns, drinking champagne under black, glistening umbrellas. It looked more like a wedding than a funeral. These simple, anonymous roadside shrines, with their plastic trucks and teddy bears, tugged more at my heart.

BECAUSE I WAS FACING the back of the ambulance, I didn't see Plaza Huincul until we were in it. One minute we were driving through a shimmering desert, and then suddenly on either side of the road there were pastel adobe houses that looked like cubes of sugar candy, and then we were passing pastry shops and furniture stores, and then we were at the museum. I climbed stiffly out of the ambulance and followed the others into a small, single-storeyed building painted a chalky yellow, with the words Museo Carmen Funes in iron letters on the front. The rain had pooled on the sidewalk by the entrance. "Don't forget to wipe your feet," Eva said at the front door. The foyer floor was highly polished, and a towel had been spread out for our boots. Two women smiled greetings from behind the counter as we entered.

Rodolfo was obviously proud of his museum. He'd become the provincial paleontologist ten years before, and the museum was his chief responsibility, after finding new dinosaurs to put in it. "When I first came here," he said as he showed us around, "the museum was in another building, much smaller, very dingy, filled mostly with modern stuff about local history. My office was the men's washroom." At that time, Plaza Huincul's reputation as one of the world's paleontological hotspots was still in the future. A few bits and pieces had been turned in by local goat farmers and hikers and were lying around in a tool shed somewhere, or else were out on loan to another museum that had a paleontologist on staff, but Plaza Huincul had nothing that required a building or even a whole room to itself. Then dinosaurs started leaping out of the ground. Rodolfo took us through a darkened passageway with backlit display cases set into the walls, and as we turned a corner we were met by one of them, a full-sized skeleton of *Giganotosaurus carolinii*, a huge theropod, the hugest, in fact, known to science. Its bony presence filled half the room. Its slavering rictus loomed over the doorway. Laid out on the floor beside it, like a fresh kill, were the bones of a giant sauropod, discovered in 1988 in a valley a couple of miles northeast of town by two kids on dirt bikes. One of its vertebrae was the size of a fifty-gallon oil drum. A photographic chart on the wall behind it identified it as *Argentinosaurus huinculensis*, "*el mas grande del mundo*," the biggest in the world.

Rodolfo's work—following up on local reports, making his own prospecting trips, training assistants to work independently while he was away on field trips—had paleontologists talking in hushed tones about "the Neuquén fauna," the amazing and diverse assemblage of dinosaurs, mostly from the Late Cretaceous and all found within a short radius of Plaza Huincul. And mostly theropods. A small, glass case held the skeleton of the ornithiscian dinosaur *Gasparinisaura cincasaltensis*, which was about the size of a greyhound. There was another, smaller theropod, *Velocisaurus unicus*, described in 1991 by José Bonaparte, Rodolfo's mentor. That same year Rodolfo found his first *Carnotaurus*, another two years later, and, as I'd read in the newspaper, a third a few months ago.

As Phil said, we were clearly in theropod heaven.

The *Giganotosaurus* was suspended from wires attached to large meat hooks screwed into the ceiling. Water dripped through the holes made by the hooks, ran down the dinosaur's ribcage and pooled on the floor by its reconstructed feet.

"Or maybe theropod hell," Rodolfo sighed. He wanted a new building. "Every time it rains this place leaks," he said. "The city has promised to make a new addition to this one, but not for dinosaurs. They say there is more to Plaza Huincul than dinosaurs, and we should have more local history displays in here. I pointed out that when we had local history displays, hardly anyone came to see them. Since I started filling up the place with dinosaurs, attendance has increased exponentially. We're getting forty or fifty people a day coming here. They're not coming to see a bunch of old typewriters from the oil company's payroll office, they're coming to see dinosaurs." He looked up at the ceiling and closed his eyes, as though he were in a church. Then he opened them again. "Last year, I went to Buenos Aires on vacation to see my family, and while I was gone the mayor told the people working here to start charging admission. This was something I had opposed from the beginning. The people of Plaza Huincul already pay for this museum with their taxes, I told him. When I got back and found out what he'd done, I went to his office and had a talk with him. I've been told my voice could be heard all over Plaza Huincul. We reached a compromise. Now we do not charge local people who come here, only those who come from out of town."

The official part of the tour over, Phil, Eva and Rodolfo went into the back room where Rodolfo kept the fossil collection while the rest of us wandered around looking at the other displays. Paul, always interested in things that didn't have backbones, was absorbed in the museum's fossil spider, a fourteen-inch megarachnid from the Carboniferous Period, which made it 280 million years old. The fourteen inches did not include the legs. J.-P. was looking at minerals. I went through a closed door and found myself in what appeared to be the local history room. At first it looked like a place where junk was kept between garage sales. The only light came through a row of thick-plated windows high along one wall, and there was the dusty smell of neglect in the darkened air. I closed the door softly and tried to piece together a human history from the objects piled on the floor, on metal tables, hung from the walls. Plaza Huincul, I learned, was really two towns that had grown together, a kind of cell division in reverse. One was called Cutral Có. Well, actually Cutral Có was the name of the oil company that had built the town in the 1920s, and became the name of the district where the company executives lived. Plaza Huincul was the worker town, the service town, where the labourers' houses were built close to the truck repair shops and gas stations where they worked. On the wall beside the door was a sepia photograph of a thin man in three-quarter profile with a high collar, a pompadour haircut and a bandit's moustache: he was identified as Enrique Martín Hermitte, Padre del Petroleo en la Argentina. More photographs of Plaza Huincul: the great flood of 1952; "An unusual fall of snow, 1928"; the first oil rig, erected 1917. A black Bakelite telephone. A Model-3444 Tube Analyser. Something called a Retromax, which I thought would be an excellent name for a time machine. A bass violin, flaking and unstrung. A cracked leather bellows. A kerosene lantern. An adding machine, three hat blocks and a lady's sidesaddle. A dentist's chair, a school desk and a Naumann treadle sewing machine. Three car-racing trophies. A cash register and a Wang computer. I began to see what the city fathers meant when they pressed Rodolfo to pay some attention to this room. This was their past, more real to them than dinosaur bones; it was something they could touch, could almost remember, flotsam salvaged in a backwater of time's flux, like suitcases hauled up from an old wreck.

I left the room by a second door that led down a hallway, at the end of which I could see the others sitting and standing in Rodolfo's office. Although it was not meant to hold so many people, at least it was not a men's washroom. The walls were lined with grey metal shelves stuffed with books, scholarly journals and the occasional bone. An olive branch protruded from one shelf—last week had been Palm Sunday, and in Argentina people take olive branches to church instead of palm leaves—and on the floor under one set of shelves was a fibula from some dinosaur or other. Rodolfo sat at a grey metal desk smoking a cigarette.

"The road into camp is impassable in this rain," he was saying. "We have to drive more than five kilometres along a dried riverbed, and when it rains even a little the riverbed fills up with water." Apparently road construction crews had been taking loads of gravel from the riverbed for years, and when the holes they created filled with water they became traps of quicksand. "Last year," Rodolfo said, "we nearly lost a whole truck in one of those pits." Also, the dirt track leading to the riverbed turned to gumbo whenever it rained. "I think we should stay in Plaza Huincul until the rain stops," Rodolfo said. "There's a lot to do here at the museum."

The rest of us agreed, and Rodolfo called a hotel to reserve three rooms. It was true that there were things we could do at the museum. Paul and J.-P. could read geological reports of the area in which we would be working. I could check out the scholarly literature for descriptions of the dinosaurs that Rodolfo and others had found in the foothills around Plaza Huincul. Phil and Eva, who had worked at the same site last year, could examine the bones they had taken out and stored in Rodolfo's collection room. Each bone or fragment of bone had to be recorded, numbered and photographed. There was enough to keep us busy for days, if the rain held out that long, although we hoped it would not.

WHILE THE OTHERS SETTLED down to their various tasks, I returned to my solitary exploration of the museum. In a room off the main corridor I found a small natural history exhibit, consisting primarily of some bleached bird skulls, a few stuffed mammals, rows of beetles and butterflies pinned to black velvet, and two rather large snake skins. Most of the specimens were local variations of familiar animals. A canine resembling a North American coyote, for example, was labelled *Zorro colorado*, red fox. But some were completely new to me. One, called a *pudu pudu*, looked like a small, hoofed dog, and there was a small armadillo known as the *piche*.

A line of skulls was arranged in a glass-topped case in order of their size: *cormorán* (cormorant), *garza* (heron), *nandú* (rhea). Of these I looked at the rhea skull the longest, and then went back to the taxidermy case where there were two stuffed rheas, one called NANDÚ and the other CHOIQUE. I had to look these up in *Birds of Argentina and Uruguay*, where they were listed as local names for the greater and lesser rhea,[3] respectively, members of the ostrich family found only in South America. The *nandú* was common in the northern, more tropical parts of Argentina, and the *choique* ranged from central Patagonia to Tierra del Fuego and the Malvinas. Plaza Huincul fell between the two zones, which meant either I could see both of them or neither. Since the extinction of the great auk in 1844 there have been no flightless birds in North

---

3 The name rhea is taken from Roman mythology: Rhea Silvia was the mother of Romulus and Remus, the twin founders of Rome, whose father was Mars. This makes absolutely no sense to me. Maybe the birds' panicked flight from bolo-swinging gauchos reminded someone of Rhea's unsuccessful flight from Mars.

America; I had never seen a flightless bird in the wild, and I very much wanted to. These particular members of the ostrich family stood about four feet high, the greater was a bit taller, but both were giant, thick-necked birds with reddish grey backs, white undersides and long, powerful legs. They had three toes, each ending in a sharp claw. Darwin had been fascinated by them on the coast. He named them "*Struthio rhea*, the South American ostrich."[4] The female, he noted in his journal, laid enormous numbers of eggs. "Out of the four nests which I saw, three contained twenty-two eggs each, and the fourth twenty-seven." He repeated what he heard from his gaucho informants, that several females shared a single nest, in other words that the twenty-two eggs in one nest might have been laid on the same day by four or five different females. They then were incubated solely by the males. He calculated that if each female were capable of laying twenty or so eggs in a season, but in clutches of four or five laid several days apart in different nests, then each multi-female nest would hold as many eggs as would a nest containing all the eggs of one female. But since the eggs in a multi-female nest would all have been laid on the same day, they would hatch at the same time, resulting in fewer broken or addled eggs per nest, and easier feeding of the chicks. In Darwin's view, this represented egg-laying efficiency at its best. But he was puzzled by the fact that there were so many single eggs scattered about outside the nests, unincubated and never hatched. They were called *huachos* by the gauchos, who collected them for food. Sometimes as many as a third of all the eggs in a colony would be *huachos*. How could such efficiency and yet so much waste be combined in a single species, he wondered? He put

---

4 Nor can I guess why Darwin persisted in calling them ostriches. The ostrich (a corruption of "estriche," meaning eastern country, from which Austria is also derived) is native only to Africa and Arabia; the South American ostrich has been called a rhea since at least 1790, when John Latham listed it in his *Index ornithologica siva systema* as a separate species from the ostrich and the emu.

it down to understandable confusion arising "from the difficulty of several females associating together, and persuading an old cock to undertake the office of incubation." This is a fine, narrative solution, and is what I like about field work, which is really a form of educated speculation. By the time Darwin collected his thoughts and published *The Origin of Species*, he had dropped his idea that such wastage was the inevitable result of multi-female nesting, and put it down lamely to "imperfect instincts." I much prefer the image he evokes of domestic squabbling, of two females arguing over an egg and breaking it. Amateurs rarely graduate from this speculative phase. W. H. Hudson, for example, also observed that a large number of ostrich eggs ended up outside the nest; he thought the males kicked out the eggs so that they would rot, attract insects, and thus provide a ready supply of food for the hatchlings. This attributes an improbable degree of foresight to male ostriches, but it makes a fine story and sounds more closely observed than "imperfect instincts."

When I had asked Phil earlier which birds were important links in the chain leading from dinosaurs to birds, he had said, somewhat enigmatically, "You might look at ostriches." The family has retained several distinctly reptilian features. William Beebe, the amateur naturalist who ran the New York Zoological Society's bird house at the turn of the last century, once received a juvenile ostrich from Africa that had been so roughed up in shipment that most of its wing feathers had fallen off, revealing two distinct, reptilian claws at the tips of its wings. He took a photograph of them, which he reproduced in his 1905 book, *The Bird: Its Form and Function*: it was a shocking sight, the weak, naked wing tipped with claws. Larry Martin had said that if you saw claws in a box of fried chicken you'd be suspicious; seeing them on the wing of a modern bird was no less disturbing. They looked like mutations, genetic throwbacks, something you would see in a textbook of human birth defects. A child born with gills or a tail. Elephant-boy. Reptile-bird. But all ostriches, rheas and emus have them. They had, according to Huxley, inherited them from *Archaeopteryx*.

OSTRICHES DO NOT, so far as I know, hide their heads in sand, but neither do they fly from danger. It is thought that they once flew, but subsequently lost the desire or the need, and then the ability. They had retained feathers on their wings, but they and their tail feathers were mainly for display. The way to tell a flighted bird from a flightless one is to look at its primary wing feathers. On a flightless bird, the primaries are symmetrical, the branches on either side of the central shaft growing evenly, like the limbs of a two-dimensional Lombardy poplar. Those of a bird that flies are asymmetrical, the branches on one side of the shaft gently curved and regular, as though combed, but those on the other side bent more sharply towards the shaft, shorter and twisted, so that the feather looks like a pine tree that has been constantly buffeted on one side by wind driving in from a cold sea.

Although flightless birds are unknown in North America, some, like the wild turkey and the roadrunner, seem to prefer running to flying, and may eventually lose the unexercised option. Others may be just setting out on the road to flightlessness. Take the peculiar case of the ring-necked pheasant, for instance.

Now fairly common throughout the western plains and Canadian prairies, ring-necked pheasants were introduced from China to Montana in the 1880s by a group of hunters who had grown tired of shooting sharp-tailed grouse, ducks, magpies, crows, hawks, geese, vultures, cougars, coyotes, wolves, antelope, bears, bison and Richardson's ground squirrels. The birds did well, and by the 1920s had spread north into Alberta. Hunters liked them because they would lie low in tall grass for a long time, then suddenly explode out of cover, sometimes at the hunter's very feet, scaring the bejesus out of him and flapping a short distance before seeking cover again. This gave the hunter only a few seconds to recover his bejesus, take aim on a fast-flying target, and blast it out of the sky. Pheasants were deemed to have at least a chance of escape, and so killing them was termed "game hunting."

In recent years, however, game hunters have begun to notice that pheasants don't seem to fly as much as they used to. More and more they have taken to running, instead of flying, from an approaching sportsman, who, being a sportsman, feels reluctant to shoot a bird that is not availing itself of its minuscule chance of escape. Often the birds don't even run, they just stay hunkered under cover even when the hunter passes within a few feet of them. This makes hunting pheasants less fun than it used to be. One hunter I spoke to said he would have quit hunting pheasants years ago if his son hadn't bought him a bird dog. The dog points to a sitting pheasant that would otherwise escape the avid outdoorsman's notice. The gamesman can then walk over to the pheasant and kick it or prod it with the business end of his shotgun, or set the dog to chasing it, anything to make it fly, you see, so he can shoot the consarned thing fairly and squarely.

When I asked him why he thought pheasants would rather run than fly, he gave a well-considered answer: "I haven't seen this written up anywhere," he said, "but this is what I think. I think that pheasants have two defences: one is to run and the other is to fly. And I think some pheasants have a greater tendency to run, and others have a greater tendency to fly. Now, over the years, those that had the tendency to fly were more readily killed off by predators. You notice, for example, that there are more pheasants flying at the beginning of hunting season than at the end of it. This would leave more birds with a tendency to run to reproduce and pass that tendency on to their offspring. Gradually, you have a predominance of runners, which is what we have now."

That was about as succinct a summation of Darwin's theory of evolution by natural selection as I'd heard. Over many generations of such selection for flightlessness, the ring-necked pheasant population in Montana and Alberta would become sufficiently different from the parent population in China that they would cease to be ring-necked pheasants at all. They would become flightless. Local hunters might take to calling them prairie roadrunners. Their breast meat would become whiter and their drumsticks darker, for flightlessness brings with it a host of attendant physiological changes: stronger and longer legs, smaller and weaker flight muscles, smaller wings, maybe larger eyes and a quicker brain, maybe fewer and larger eggs. This is how Dale Russell thought about the future of *Troodon*. Physiological changes eventually lead to an altered social structure, modified to ensure the survival of offspring, perhaps involving increased cooperation among females. Hunters in Montana and Alberta may well be witnessing ring-necked pheasants in the process of becoming as flightless as ostriches.

OSTRICHES SEEM TO EXIST to challenge our notion of what birds are. They are, for example, too big to be proper birds. Sir Thomas Browne, the great seventeenth-century English physician and natural historian who, according to Coleridge, had "a feeling heart and an active curiosity, which, however, too often degenerates into a hunting after curiosities," kept an ostrich in his garden. The bird, Browne wrote, was "either exceeding or answerable unto the stature of the great porter unto King Charles the first." That put it over seven feet high, well at the extreme end of possibility for flighted birds. "Whosoever shall compare or consider together the Ostridge and the Tomineio or Humbird," he wrote, "not wayghing twelve graines, may easily discover under what compass or latitude the creation of birds hath been ordained." As for diet, whereas Darwin mentions simply that they were known to eat fish, Browne records an astonishingly omnivorous appetite: the ostrich in his garden "soone eat up all the gilliflowers, Tulip leaves, & fed greedily upon what was green, as Lettuce, Endive, Sorrell: it would feed upon oates, barley, pease, beanes, swallow onyons, eat sheepes lights and livers." When it swallowed an onion, Browne notes, the bally thing became stuck in its throat so that its long neck resembled a snake swallowing a monkey.

Huxley also wrote about ostriches. It was his comparison of an ostrich skeleton with that of the huge theropod *Megalosaurus* that first led him to surmise that birds were descended from dinosaurs. Of all the myriad types of birds in the world, he thought, ostriches were the closest to dinosaurs. They were long-necked, long-legged, short-armed and flightless. He thought they'd always been flightless. His theory of bird evolution placed the ratites—flightless birds such as ostriches, kiwis, emus, rheas and the extinct New Zealand moa— in a direct line of descent from small, bipedal dinosaurs, in much the same way that human beings were in a direct line of descent from apes, with flying birds more evolved than flightless ones. "The road from Reptiles to Birds," he wrote in 1868, "is by way of Dinosauria to the ratitae. The bird 'phylum' was struthious. . . ." In other words, the line of descent went from primitive reptiles to dinosaurs to ratites to songbirds.

When *Archaeopteryx* was found in the Jurassic slate beds of Solnhofen, however, it put a severe kink in Huxley's line. The old-est fossil bird clearly had a wishbone and flight feathers: *Archaeopteryx* could fly. It was therefore, if Huxley's theory was correct, more advanced than a flightless ostrich, which could not be the case. If *Archaeopteryx* had evolved from dinosaurs, then ostriches must represent some other, earlier branching off from the main line—a branch that did not lead to modern birds.

There was more. *Archaeopteryx* had been found in the same bed, and was therefore the same age, as the indisputably non-avian dinosaur *Compsognathus*. And *Compsognathus* was almost identical to *Archaeopteryx* in everything except for the troubling matter of wishbone and feathers. Both were small, chicken-sized, toothed and long-tailed. If *Archaeopteryx* had not had feathers, it would easily be mistaken for *Compsognathus*—in fact, one specimen *was* misidentified as *Compsognathus* and remained so for many years. If Darwin were right, evolution from dinosaur to bird ought to have taken countless generations, hundreds of minuscule adjustments, perhaps millions of years. And yet here was a bird and the dinosaur it had probably evolved from lying beside one another in the same bed. Darwin had an irritating (to Huxley) habit of repeating that "*Natura non facit saltum*," nature does not make leaps. But in Solnhofen the fossil record showed a definite leap: dinosaur one day, bird the next (or maybe the other way around?). Huxley wondered idly if perhaps *Compsognathus* hadn't been feathered, too? Perhaps it was really a flightless bird, in which case finding *Archaeopteryx* and *Compsognathus* in the same bed would be little more startling than finding a penguin and a cormorant together? On the other hand, perhaps *Archaeopteryx* was simply off the main path from dinosaurs to birds, a diversionary sideroad, one of nature's experiments that for some reason was not followed up for a few million years? Or perhaps it was simply not true that nature never made leaps. "We believe," Huxley wrote in 1860, "that nature does make jumps now and then, and a recognition of the fact is of no small importance in disposing of many minor objections to the doctrine of transmutation." This was a very unDarwinian way of explaining Darwinism.

But it showed that Huxley was contemplating three revolutionary scientific ideas long before there was much hard evidence to support them. He proposed that birds evolved from dinosaurs, that evolution might proceed in sudden leaps—a theory of evolution that did not come into general currency until Stephen Jay Gould

and Niles Eldridge brought it out and dusted it off in the 1980s—
and that some dinosaurs might have been feathered. Eventually, the
fossil record would show that, in all three cases, Huxley was on the
right track at the wrong time.

CHAPTER FIVE

Desert Rain

I T RAINED FOR THREE days and three nights. We took rooms in the Tortoricci, a small, dimly lit hotel in Plaza Huincul, and spent our days at the museum and our nights reading or watching television. I shared a room with Mike Getty. He had been on the road for two months already, having spent six weeks hitchhiking through Chile with his girlfriend, sleeping in tents and caves, before coming on to Argentina while she returned to her job at the Calgary zoo, and so he was perfectly content to hang around Plaza Huincul before leaping into another adventure involving foam mattresses and open-fire cooking. Lying on our bunks, listening to the swish of cars on the wet street outside our window, we watched movies (in English with Spanish subtitles) and read before falling asleep. The movies blurred into one another. I remember *GI Jane* because a lot of it took place in the rain. Most of the films involved loud explosions and hand-to-hand combat. Somebody saved the president's airplane from being hijacked, somebody else frustrated a plot to kidnap the Chinese ambassador's son.

At the museum, Phil and Rodolfo were busy comparing the dinosaur bones they had uncovered the previous year with the corresponding bones from *Giganotosaurus carolinii*. Rodolfo didn't know what species the new bones belonged to, but they were close enough to *Giganotosaurus* to make the comparison interesting.

There was a sand table in Rodolfo's lab, and we arranged the bones side by side on it: *Giganotosaurus* vertebra, new species vertebra; *Giganotosaurus* dentary, new species dentary. With the sand holding them in place, Mike photographed each pair from four different angles—lateral, anterior, distal and proximal views—using Thom Holmes's black jacket as a backdrop. Thom was a writer from Philadelphia who was working on a series of children's books about dinosaurs and wanted some first-hand field experience, which he hoped he would soon be getting. He was staying for only ten days, and already three of them had been eaten up with rain. He was boarding with Rodolfo and looking a little bleary-eyed. The museum was open until nine o'clock, after which he and Rodolfo and Rodolfo's wife, Claudia, and their daughter usually went out for dinner. "Argentines don't even think about food until ten o'clock at night," he complained, watching us get sand all over his jacket. "It's usually after midnight before we get home."

Phil had arranged the new species' metatarsals, the long, spindly bones connecting the ankle and toes, on another table and was pondering their peculiarities. They were huge, for one thing, each about the size of a human leg bone. And there were too many of them, enough for three or four left feet. No *Giganotosaurus* metatarsals had ever been found, so there was nothing to compare these with, but it was obvious already that this new, unnamed species was a theropod almost as big as *Giganotosaurus*, and closely related to it. "Same genus," Phil said at one point, "but definitely a new species." Not *Giganotosaurus carolinii*, then, but probably *Giganotosaurus something-or-other*.

Twice a day Susanna, one of the museum's volunteers, brought us maté. Maté is gaucho tea, sometimes known as Jesuit tea because, in the seventeenth and eighteenth centuries, when the Jesuits ran Paraguay much as they ran New France, they became so wealthy and powerful that the Vatican suspected them of having found El

Dorado; in fact, they financed their New World ventures by raising and selling maté.

It is made by placing the crushed leaves of the yerba plant, *Ilex paraguariensis*, a variety of holly that is rich in caffeine, in a small, polished gourd, adding hot water to it and drinking the result through a *bombilla*, a thin, metal straw with a strainer at the bottom. Like holly in Europe, maté is a plant of mythological proportions, in cultural terms a sort of cross between tea and coca leaves. One holds the gourd like the bowl of a pipe, and sucks through the *bombilla* with a deeply contemplative expression, like Sherlock Holmes pondering a three-pipe problem. Everything stops for maté. You could certainly halt a revolution in rural Argentina by running out into the plaza and shouting, "Time for maté!"

There are many stories concerning the origin of maté. One has it that the yerba plant came from a place near the village of Tacurú Pucú, in Paraguay, where an old man had moved deep in the forest to protect his beautiful daughter from men and other evils of civilization. One day, God came down to Earth and, along with St. John and St. Peter, travelled around South America. They stopped for the night at the old man's lodging, and the old man welcomed them, although of course without recognizing them, and even killed his only hen for their supper. After eating, the three holy travellers stepped outside, and God asked St. John and St. Peter whether they thought He should reward the old man for his generosity. St. John and St. Peter said they thought He should, and so God called the old man out and said to him: "Thou who art poor hast been generous, and I will reward thee for it. Thou hast a daughter who is pure and innocent, and whom thou greatly lovest. I will make her immortal, and she shall never disappear from Earth." And so saying, God instantly transformed the old man's daughter into the yerba plant, so that no matter how often men cut her, she would grow again.

This is supposed to be a happy story, but to me, a father of two daughters, it was the kind of tale that Annie Dillard calls a "God-fearing explanation of natural calamity, harsh all around." Far from making me think highly of the yerba plant, I found myself wondering what kind of culture would honour such a cruel and thoughtless God, a God who would take a man's daughter from him and turn her into a national drink, before his very eyes, before he could even say goodbye, and condemn him to watching her fall to the harvesters' knives season after season, year after year. Even if he believed that the plant was his daughter—let's say he did—what consolation was that? A plant was not a child. Besides, plants were not immortal. I imagined the old man's constant anxiety, his fear of insects and fungus and rot, of too little or too much rain, of the harvester's axe, as he waited to see if the ravaged stubble of last year's yerba crop would sprout again.

WHILE THE RAIN DOGGED us and kept us in Plaza Huincul, I tried to get to know Rodolfo better. As I had suspected from the newspaper story about the *Carnotaurus* discovery, he was what, if I were a reporter, I would describe as friendly but not forthcoming, somewhere between effusive and elusive. He would answer direct questions, but rarely volunteered any information concerning his work. I couldn't figure him out. He was nervous around writers, he said, portraying himself as a simple country boy who had suddenly found himself in the big time. "I'm just a digger," he said. But there was more to it than that. The crush of reporters that had called him after the *Carnotaurus* showed up had made him wary, or perhaps just weary, of interviews. "What *Carnotaurus*?" he said when I asked him about it. "We don't know if it's a *Carnotaurus* because there is no skull." He even told me he had been instructed by the government not to speak to the media, to refer all further requests for information to the provincial public-relations department. He and Luis Chiappe had an agreement with the *Los Angeles Times*, which broke the story of the new *Carnotaurus*, not to let any other reporters see the dig site. I told him I wasn't a reporter, I was a writer, but he didn't seem to be reassured by that. I suspected that he tolerated me mostly because he was Phil's friend, and Phil had invited me. To Rodolfo, friendship was everything.

He was born in a small town in Patagonia, but his family moved to Buenos Aires when he was two years old. Like many Argentines, he was of mixed Italian and Spanish background. His father did something in media marketing and advertising, which may explain why Rodolfo didn't trust reporters. He was one of those kids whose interest in science preceded puberty; in Rodolfo's case, it was biology. He loved animals. He enrolled in veterinary school after high school, but the idea of spending his time looking after other people's pets put him off. He wasn't interested in curing animals, he told me, but rather in figuring out how they functioned, in their anatomy. For a while (like Che Guevara) he considered moving to Patagonia to work with cattle breeders, but, as he said, he was "more interested in dead animals than living ones," which was not a high recommendation in a vet. So he went to teachers' college to become a science teacher, with an emphasis on natural history and comparative anatomy. He was already very good at comparative anatomy. "We used to make a game in class," he said. "We would form partners, and one partner would toss a bone across the room to the other, and the second partner had to identify the bone before he caught it. As the year progressed, we would move closer and closer together, until by the end of term we could identify any bone from any animal practically on sight."

In the end, he never looked for a teaching job. While still at teachers' college, he began working as a volunteer at the Buenos Aires Museum of Natural Science, with José Bonaparte. When Rodolfo met him, Bonaparte was the most famous paleontologist in Argentina, having found and named half a dozen dinosaur species already. Rodolfo described him as a kind man with "a special kind of light in his eyes," and the light shone on the eager young anatomist. "He was easygoing, interested in everything, and he showed me that it was possible to make a living as a paleontologist."

Rodolfo divided his time at the museum between technical work—preparing and identifying bones collected in the field by Bonaparte's team of volunteers—and drawing the illustrations for Bonaparte's numerous scientific papers. To his surprise he became so good at both that in 1985, when he was twenty-three, he was taken on by the museum as a permanent, salaried employee. To his further astonishment, he also became its curator of vertebrate fossils. "Most of the collection was mammals," he said, a little disdainfully.

For a century and half, Argentina had been known not for dinosaurs but for its weird mammal fossils. The first of these were noted by Darwin in 1830, during his voyage on the *Beagle*: the area around Bahia Blanca, he wrote, was "highly interesting from the number and extraordinary character of the remains of gigantic land-animals embedded in it." In 1887, Carlos Ameghino collected what were probably the very bones Darwin had seen. Ameghino's older brother, Florentino, then vice-director of the new Museo de La Plata, had a theory that Argentina was the birthplace of all the mammals, including human beings, that had since dispersed throughout the world, and he sent his brother Carlos to Patagonia to find evidence for it. At the mouth of the Rio Negro, Carlos turned up a rich horde of fossil mammals. "Full of joy," he wrote to his brother, "I return to our camp forgetting for a time the fatigue and the labour, ready to face others perhaps more severe in the future."

Darwin didn't even mention dinosaurs when he was in Patagonia. He listed nine mammals[1] that he found lying together in a *barranca* rising from the beach, all within an area of about two hundred square yards. South America "must have swarmed with great monsters," he wrote, "but now we find only the tapir, guanaco, armadillo, and the capybara, mere pygmies compared to the antecedent races." The antecedent races: he was already thinking in terms of what he called at the time "the law of the succession of types." Paleontologists for the next hundred years would think of Patagonia chiefly in terms of mammals. When George Gaylord Simpson tramped through Patagonia in the 1930s, a century after Darwin, he too found the bizarre forms of South America's pre-Pliocene mammals "rather startling." They didn't look like anything else. "If you dig bones in North America," Simpson wrote, "you find ancient elephants, three-toed horses, rhinoceroses with paired horns, saber-toothed cats, and the like." Ordinary stuff, animals with known counterparts in the modern world. Even a giant sloth is just an ordinary sloth blown up. We can imagine a horse with toes instead of hooves, in the manner of a Magritte painting. In Argentina, however, Simpson found animals with no living analogs: "astrapotheres, and homalodontotheres, and sparassodonts, and toxodonts, and pyrotheres," creatures that looked like rather ridiculous illustrations to children's stories. *Sesame Street's* Mr. Snuffalupagus, which looks like a cross between an elephant, a bloodhound and a musk ox, might have been a South American mammal. The astrapothere, for example, was roughly the size and

---

1 Including the Mylodon, whose supposed skin in his grandmother's cabinet so famously fired Bruce Chatwin's imagination. "The beast of my childhood dreams," Chatwin wrote later, "turned out, in fact, to be the mylodon, or giant sloth—an animal that died out in Patagonia about 10,000 years ago, but whose skin, bones and excrement were found, conserved by dryness and salt, in a cave at Last Hope Sound, in the Chilean province of Megallanes." The original of this remarkable specimen was supposed to have been in the British Museum, but W. P. Pycraft makes no mention of it in his 1905 history of the museum. He does, however, mention a creature very similar to the Mylodon: "the Grypotherium, whose remains, marvellously well preserved, were found only a year or two ago in a cave in Patagonia." Pycraft says that the Grypotherium bones "present a remarkably fresh appearance, retaining shrivelled remains of gristle and flesh," and were found with "large lumps of dung." Could Chatwin's grandmother's cabinet have really contained bits of a Grypotherium, not a Mylodon?

shape of a rhinoceros, but had an anteater's nose and the tusks of a wild boar. Its lower teeth, wrote Simpson, "would make elegant lodge emblems"; they were large and scalloped at the ends, like sheaves of wheat.

The cause of all this idiosyncrasy was the total isolation of South America from the rest of the world while these animals were evolving: the Isthmus of Panama, known to geographers as a "filter bridge," was not there at the end of the Cretaceous Period. When mammal evolution started taking off after the disappearance of the dinosaurs, South America was an island continent, like Australia (which still has strange animals, the marsupials). Animals from North America, which could wander over land bridges and intermingle with local species or adapt to new environments, could not get to South America[2], and so South American mammals evolved on their own into weird and wonderful creatures. Then, when the isthmus formed during the Pliocene Period, about six to nine million years ago, North American animals began funnelling down into South America, and vice versa. These *norteamericanos* were tougher and more efficient; many of them had already learned the art of self-defence by having faced invaders from Europe and Asia; thus the camel, chased out of North America by the horse, supplanted the South American litopterns (hoofed creatures with long, anteater-like snouts) and evolved into llamas. By three million years ago, most of South America's bizarre native forms had become extinct.

Under Bonaparte, Rodolfo was getting a crash course in early South American dinosaur dispersal. And what he and the other members of Bonaparte's team—Jaime Powell, Fernando Novas,

---

2 There are some exceptions to this, as to every, rule: rodents, monkeys and raccoons seem to have turned up in South America before the Isthmus of Panama linked the two continents together. Being relatively small animals, they probably drifted south on massive seaweed rafts, possibly working their way from one Caribbean island to the next until they hit South America. The larger animals had to wait for the Panamanian causeway.

Luis Chiappe—were learning was changing the paleontological world's notion of South America's place in the history of evolution. In January 1989, when Bonaparte was excavating the sauropod *Argentinosaurus*, the newspapers were suddenly full of news that the world's largest dinosaur had been discovered in Neuquén Province. The museum in Plaza Huincul wanted that dinosaur. Bonaparte refused to give it to them. It was, he said, too important a specimen to be hidden away in a tiny museum in a remote province of Patagonia that didn't even have its own paleontologist to take care of it. Plaza Huincul said it would hire a paleontologist.

Rodolfo had already applied for a job at two other institutions in Patagonia without success. He didn't know why he had been turned down, but presumably Bonaparte was not sorry that he had been, since Rodolfo was a valuable assistant to keep in Buenos Aires. In Rodolfo's version of the story, it was Bonaparte who suggested that Plaza Huincul hire Rodolfo: "When I joined Bonaparte's field team in February," he said, "Bonaparte said to me, 'I've got a job for you in Plaza Huincul.' 'Great,' I said. 'Where's Plaza Huincul?'" Whether or not this version is true, the Plaza Huincul job was perfect for Rodolfo. The city agreed to provide him with a house, and to hire his wife Claudia as a teacher. In fact, Rodolfo thought at the time that the council was more interested in Claudia, who is now the city's top special-education teacher, than they were in having a paleontologist forced on them, especially one who was soon going to pressure them into providing a newer and bigger building for dinosaurs. But Bonaparte was fortunate to have a loyal colleague in Plaza Huincul, presiding over his precious *Argentinosaurus*: when Bonaparte's paper describing the specimen appeared in 1993, Rodolfo's name was on it as sole co-author. Rodolfo was equally fortunate. *Argentinosaurus* drew the world's attention to Neuquén Province as a rich source of controversial new dinosaurs, and put Plaza Huincul at the centre of one of the most fossiliferous dinosaur fields in the world.

ON SUNDAY MORNING I woke at seven a.m. thinking something was wrong. It took me a moment to realize what it was: the hotel was too quiet. The rain had stopped during the night. I got up, dressed and went outside. It was still slightly dark and the sky was overcast, but I could see stars in the west, and the puddles on the sidewalk were undimpled by fresh rain. Instead of turning left on Avenida Rodriguez towards the museum, I walked in the opposite direction, towards town, breathing in the scented air and looking about with the benevolent eye of farewell. The shops were closed, but people were beginning to stir. I passed a computer store, a health spa ("*Masages de relax*") and something called Red Informaciones Inc. Down the centre of the avenida was a wide boulevard with a sidewalk in the middle lined with low birch trees in which birds were stirring, and I crossed over to it. It was now light enough that I could see open country at the far end of the street, over the crest of a small hill, and I walked for an hour without reaching it, past fruit stores and newspaper kiosks, a dollar store ("$1.99 y *mas!*"), and eventually a residential area where schoolchildren, all wearing white smocks over their regular clothes, were massing at the corners before flocking off to school. Thirty minutes later I finally reached the end of the avenue; it stopped abruptly at the entrance to a large field of grass-covered sand and low shrubs, where invisible birds burdened the branches with chirping. Beyond the field the open steppes rose in a succession of steep hills as though they'd been piled there to make room for the town, each hill a different tone of brown and grey.

WE WERE PACKED and ready to go by noon. Three vehicles, all of us jammed in somehow, with our camping gear and quarry tools and enough food and water for several weeks stacked high in the back. We had been joined the previous night by an American writer, Don Lessem, and his partner, Valerie Jones, and so there were fifteen of us altogether. We stopped once for gasoline and beer, then ground our way out of town along a roughly paved road into the salal desert. Being in it was infinitely better than seeing it from the end of the boulevard. The greens were vibrant after the rain. The steppes receive 150 millimetres of rain during the year, but we seemed to have had it all at one go; now the cold nights would set in to lock the moisture in the ground. Thorn in Spanish is *espiño*, but these thorny shrubs had indecipherable local names: *calafate, colapiche, quilembai, farillas*. Spiked grasses called *coirónes* shot up from central clumps like cartoons of exploding land mines. According to my bird book, I could expect to see carbonated sierra-finches and rusty-backed monjitas, the elegant-crested tinamou and the chocolate-vented tyrant. I studied the illustrations in the book carefully and looked out at the shrubs whirling by; there were birds out there, but we were moving too fast. The road passed directly through the town dump, acres of garbage scattered on both sides, a million white plastic bags caught on the thorns, with packs of feral dogs and pairs of aged horses rummaging among them. As we passed, huge, floppy birds stirred up irritably and settled back down; they looked to me like brown ravens with white patches on their wings: they were crested caracaras, known locally as *caranchos*. I had read about them in Hudson and Darwin; Hudson admired their grace and ingenuity, but to Darwin, who was faintly disgusted by their scavenging habits, they were a blight on the landscape. Groups of them, he noted in his *Journal*, would roost along the roadside like vultures (Darwin hated vultures, one of the things about him that infuriated Hudson), waiting "to devour the carcasses of the exhausted animals which chanced to perish from fatigue and thirst." They would "attempt to pick the scabs from the sore backs of the horses and mules," he added with a shudder, and were known to attack and kill small or wounded animals. "A person will discover the *Necrophagous* habits of the Carrancha," he wrote, "by walking out on one of the desolate plains," as we were about to do, "and there lying down to sleep. When he awakes, he will see, on each surrounding hillock, one of these birds patiently watching him with an evil eye."

After half an hour we turned off the main road onto a dirt track that twisted across the desert among huge thorn bushes that seemed to reach out and claw at our roof tops. We quickly closed the side windows. At one point, when we stopped at a place where the track was still filled in with water, I rolled down a window and cut a thorn off with my knife. It was bright green, like a coiled leaf, four inches long, thick as my finger at one end and sharp as a talon at the other. The bush itself was without leaves, but bristled with these murderous thorns. As there was nothing around (except me) that seemed to want to harm it, I wondered what it was protecting itself from.

The mud on the track turned out to be navigable, and we ploughed on at a frightening speed until we reached the edge of a low cliff overlooking a wide, flat riverbed. Not a drop of water flowed in this channel, but we could see from the wave patterns on its sandy bottom and the deep cuts into its banks that a lot of water recently had. The track led to a small cutbank, and to my consternation we zoomed down the embankment and spun out onto the riverbed, the treacherous Cañadon de Allambe that had so worried Rodolfo three days before, without so much as stopping to see if the bottom was firm. Fortunately, it held. We rolled down our windows and continued along this natural highway for another three or four miles. Everywhere the great sand cliffs loomed above our vehicles, pocked by nests of cliff swallows and knitted in place by the snaking roots of thorn bushes. Above the cliffs the rounded hills rose like piled skulls, but the sky was clear and blue, and our spirits improved with each mile gained.

Rodolfo pulled the lead vehicle up at a spot where a smaller canyon, narrow and deep, joined the riverbed, and we all tumbled out. This canyon was too narrow to admit the trucks, and so we had to go the rest of the way in on foot. When the combined loads of all three trucks were spread out on the sand, we began packing it up the canyon to the place where Rodolfo and Phil had camped last year. The temperature had risen steadily all morning, and now we could see tendrils of heat rising from the distant sand, although it was still damp and cool in the canyon. I staggered along for several hundred yards between high, sandstone walls riddled through with cliff-swallow holes, although I could not see any swallows or nests inside them. The holes were deep and curved, and many of them connected within the cliff, so that in places the solid sand seemed ready to collapse from being so honeycombed by holes. On the sandy floor were the tracks of numerous small animals, mostly goats. Mike, walking ahead of me, pointed out the tiny, delicately clawed prints of a small armadillo, the *piche* I had seen in the museum. After a few hundred yards we came to a steep path that took us up the canyon wall and along the top of the cliff, into the heat and still heading inland. I wished I hadn't brought quite so many books. We followed a sort of ledge of flat land at the base of high, rocky hills between which the canyon had been carved by years of flash floods and wind, which whistled through the swallow holes beneath our feet in a pleasant but ominous manner. The path wound among thorn bushes for perhaps another two hundred yards, and stopped at a pile of large boulders surrounding what was obviously a well-used fire pit with an iron grate over it. Here Christian and Daniel dropped the kitchen tent, and Rodolfo, sighing deeply, said: "Home at last."

I continued along the top of the canyon to where it became narrow enough that I could leap over it, and set up my tent on the other side, at a point where the canyon was joined by an even smaller one that wound down between the hills across from the fire pit. Thom, J.-P. and Paul were also in this area, each slightly hidden from the others by high thorn bushes. Rodolfo and his three technicians were set up between us and the kitchen tent. I couldn't see Phil and Eva's tent, or Mike's, but guessed they were behind a hill that rose above the fire pit. I unpacked my gear, stacking the books on a plastic sheet along one side of the tent and spreading out my sleeping bag along the other. There was not much room between them, but I put my clothes in one corner and my boots in another, and the result seemed liveable. After we had all made similar domestic arrange-ments, we congregated at the fire pit and helped set up the huge kitchen tent. Then we hiked back to the vehicles and packed in the food and cooking supplies. By the time everything was set up it was getting dark, which in Argentina meant it was almost time to start thinking about dinner.

"First we have to gather some firewood," Rodolfo said. This seemed, at first, a curious statement, for there was not a tree any-where in sight. Nothing bigger than a thorn bush, the tallest of which was maybe six feet, and none more than an inch or two in diameter at the base. But as Daniel and Christian showed us, most of the bushes grew from a cluster of denser wood at their base, grey, twisted trunks stunted by too much sun and not enough water, much like the taiga that grows in the Canadian north, which can be a hun-dred years old and no thicker than a wrist. Here the dead wood beneath the thorns was dry and brittle, and it was easy to reach in with our bare hands and break it off. Soon we were dragging great piles of it into camp. When we had stacked it neatly beside the kitchen tent, Rodolfo came over and looked pensively at it.

"You should wear gloves when picking up dead firewood," he said. "You might get spider bites."

"Spider bites?" said J.-P. "What kind of spiders?"

"Black widow spiders," said Rodolfo. To his credit, he seemed slightly embarrassed at having forgotten to mention this detail earlier. Black widows, *Latrodectus mactans*, are among the most poisonous of all spiders: the North American variety is bad enough, but the South American species is ten times more poisonous. Though it's only half an inch long, its bite is more virulent than that of a rattlesnake and its venom is strong enough to kill birds and lizards and, in ten per cent of cases, people. The poison works as a neurotoxin, and can cause muscle spasms, abdominal cramps, cellular breakdown and something the text books call "necrotic disintegration." I really did wish I hadn't brought so many books. According to my *Insects of South America*, the black widow had "the worst reputation enjoyed by any spider," and was "entitled to it." Pain caused by a bite usually showed up in the intestines three days after the event, and the bite itself was not accompanied by local swelling, so you couldn't tell when or where you'd been bitten.

"We have rather a lot in this area," Rodolfo went on. "They like to hide in dead wood. We always wear gloves. Even so," he said, looking vague, "there are many incidents."

"Incidents?" said J.-P.

"Well, deaths."

"Deaths?" J.-P. eyed the woodpile. "From black widow bites? How many deaths?"

"About one hundred per year."

"A hundred deaths a year?" J.-P.'s voice was rising. "In Argentina alone?" he said.

"No, just in Neuquén Province. One woman who was a volunteer at the museum, her father was bitten by a black widow spider. It took him twelve hours to die. I visited him at the hospital. He died a slow, painful, agonizing death." He shook his head, then added, as though he thought it might reassure us: "But she doesn't work at the museum any more."

PERHAPS I AM inclined towards mysticism, as was Alfred Russell Wallace, for instance, the amateur naturalist who came up with a theory of evolution virtually out of nowhere, in an afternoon, on the back of an envelope (metaphorically speaking), and sent his musings to Darwin just as Darwin was about to publish his own deliberations of twenty anxious years. Selection and adaptation, the twin engines of Darwinian evolution, had come to Wallace in a sort of flash, he said. His autobiography, written long after Darwin's death, makes it clear that Wallace was one of those hardworking, tireless observers who, over the centuries, have contributed immensely to the advance of science, a generous-minded duffer devoted to the hunting and gathering of data. In later life he embraced Mesmerism and was a familiar figure around the séance tables of the great mediums of England and the United States, duly noting down apparitions from Egypt and India. In Philadelphia he records calling up the image of a tall, bearded man whom he thought at first was Darwin himself, but who turned out to be a cousin of Wallace's who had recently died in Australia. I was not expecting Darwin's ghost to beckon from the foot of my sleeping bag, but I confess to thinking of a hill with a dinosaur buried in it more as a gravesite than a workplace. Many scientists feel much the same; they approach a quarry with a degree of seriousness, almost of reverence, as a believer might approach the tomb of a holy man. A dig is an exhumation. A dinosaur site is a resting place, and sleeping beside one is like pitching your tent next to an ancient cemetery; whatever ghosts it may contain have been around for a long time, and any spirit you might raise from it will be as benign as that of a distant cousin in Australia. And if you ask it the right questions, it will have as many secrets to disclose about life on the other side.

CHAPTER SIX

Overburden

THE TEMPERATURE WHEN I woke in the morning was barely above freezing. The sun made the walls of my tent almost luminous, light enough to see that during the night my breath had condensed on the inside of the nylon and was now running down onto my clothes and books. Instead of an air mattress I had brought a camp cot, a stretcher-like affair that consisted of canvas sewn between two metal poles attached to legs. Also during the night the canvas had ripped slowly away from the poles, gradually lowering me to the cold, wet floor, so that when I opened my eyes the bare poles of the cot frame loomed above me like the bars of a crib, and for a long moment I had no idea where I was. Suddenly the maniacal screams of an American rock band pulsed from a tinny-sounding radio somewhere outside. I dressed lying on my back, not an easy operation, and climbed out of bed with much difficulty. It was eight o'clock.

Down at the kitchen tent Daniel had a fire going, and a kettle of water was steaming on the grate. The others were already standing around the fire or at a folding table laid with the makings of breakfast: jars of instant coffee, a Tetra Pak of sterilized milk, a box of cereal, powdered cocoa, instant oatmeal, all the familiar American brands—Kellogg's, Nescafé, Fry's, Quaker—in peppy Spanish translation. I read the oatmeal box while I drank my first coffee.

Oatmeal was not considered a breakfast cereal in Argentina, it seemed, but rather as a thickener for things like soups and spreads. At your next dinner party, urged the beaming, bewigged gentleman on the back of the box of Quaker Instantanea, surprise (or startle) your guests with a lively (or perhaps zesty) plate of Temptation Tomatoes: bring to a boil half a cup of oatmeal, half a cup of milk, one and a half ounces of mayonnaise and a tablespoon of mustard; spoon into hollowed-out tomato halves and Hey presto! Tomates Tentación. Try it with mushroom caps, too! There was also a jar of something else I hadn't seen before, a thick, sweet, creamy spread called *dolce de leche*. It tasted a bit like cream-cheese icing for carrot cake mixed with Eagle-brand condensed milk and a chocolate hazelnut spread, only sweeter. J.-P. had already discovered it and the jar was half empty.

I joined Mike and Rodolfo at the fire, warming my hands on my coffee cup. Mike had pointed the quarry out to me the previous night, when it was dark, and now we gazed across the canyon and up the hill at it. To get to it, we would have to descend into the canyon, then climb the other side to a height of about two hundred feet, to a point where the red sandstone slope met a vertical cliff of greenish yellow rock. The cliff rose only another seven feet or so. Where the hill met it there was a slight flattened area, a sort of notch, and that was the quarry. Mike had been here with Phil and Rodolfo the year before. The notch, he said, represented six weeks of backbreaking labour. I said it looked pitifully small from down where we were standing. "It looks pitifully small from up there, too," he said.

Rodolfo said the site had been found by one of the local goat herders. "He came into the museum in Plaza Huincul with a bone, and said he had found it while looking for some lost goats. I looked at the bone; it was sauropod, I thought not very interesting, but he said there was a lot more of it, so I wrote down where he found it and didn't think much more about it." Two years later, when Phil and Eva came down to scout sites for last year's expedition, Rodolfo remembered the sauropod bone and brought them here. "We spent a few hours looking around, and realized that most of the bones coming out of the hill were theropod bones, and big ones. So we came back last year to have a more organized look, and took out about half of what we think is here. This year, we hope to get the other half."

"Do you think it's *Giganotosaurus?*"

He shrugged. "I think it's related to *Giganotosaurus*, maybe a new species. It belongs in the carcharodontosaurid family, which strengthens our links with the dinosaurs of Africa. But what is more interesting to me is to be able to get some idea of the ecology of this place as it was in the Cretaceous Period; what the environment was like, what kinds of animals were here, how they interacted. This has never been done before. Every new species we find adds to the picture."

At nine o'clock we cleared the breakfast things away and started for the quarry, all but Paul and J.-P., who went off in a different direction to look at geological features. Daniel had already carved steps leading down the canyon wall from the kitchen, and the far slope was gentle enough for us to go up at an angle without too much difficulty. About halfway up we began to see bone fragments in the loose gravel under our feet, small pieces about an inch or two long. "This is the sauropod material," said Mike as we climbed. "Pretty fragmented, although there are some bigger chunks down at the bottom. It looks like it's coming from a layer underneath the theropod stuff." We climbed on. I wondered why we hadn't made camp on the same side of the canyon as the quarry, perhaps at the top of the short cliff above the quarry itself. When we got to the top I found out. Poking my head above the cliff I was struck full in the face by a blast of wind that nearly knocked me back down the slope.

I recalled a phrase from George Gaylord Simpson's book, something about "the terrible Patagonia wind." He had been farther out on the pampas, where the wind was so strong and so constant that he had been able to walk straight down cliffs, holding his coat out like a parachute and letting the wind keep him from falling headlong into the canyon. The wind here wasn't that strong, but it was strong enough. Luckily, it was blowing from the west, from the direction of the Andes, which meant the weather would be cold but dry. On the steppes, Rodolfo said, an east wind always brought rain.

THE PREVIOUS YEAR'S quarry had exposed only a small percentage of the actual bonebed; at the end of the field season, it had been apparent that the rest of the animal disappeared farther into the hillside, and that a whole lot of overburden would have to be removed before more bone could be taken out. When a large skeleton disappears into a hill, you don't tunnel in after it. You remove the hill. The idea is to get at the bone from above, which makes for more work, but is also much safer. Paleontologists are not mining engineers; few know much about digging into cliff faces, shoring up rock roofs, reading shear factors, avoiding cave-ins. Small skeletons might be collected the easy way, but tunnelling out an animal as big as *Giganotosaurus* would have presented problems with lighting as well as structural integrity. A cave-in could put an abrupt end to a promising field season, maybe even a promising career. Even worse, it might damage the bone.

Phil and Rodolfo estimated that the bonebed was approximately twelve feet below the point where the slope met the cliff, and covered an area of about thirty square yards. That meant we would be removing more than a hundred cubic yards of solid rock just to get down to the bone level, which could be another five or six feet thick and would also have to be removed. For the next six weeks, we would be human rock-moving machines. Daniel opened a large wooden crate that he and Adrian Garrido, the third *técnico* from the museum in Plaza Huincul, had hauled up from one of the trucks. From the box, he took the tools of our trade and handed them around. Three short-handled shovels, three picks, a six-pound maul, several two-and-a-half-pound hammers and an assortment of cold chisels. Then Christian appeared over the lip of the hill pushing an ancient, rickety wheelbarrow. The wheel, long ago stripped of its rubber, wobbled drunkenly, and one of the metal handles appeared to have become detached from the dented and rusted barrow, which itself seemed badly in need of a welding job. With these primitive tools we would have to break up and cart away more than eighty tons of rock.

A new dinosaur doesn't officially exist until someone publishes a description of it in a scientific journal, which is why Rodolfo had been annoyed when the newspapers identified his latest find as a *Carnotaurus*. The original *Giganotosaurus* had simply been "an unnamed theropod from northern Patagonia" until Rodolfo described and named it in *Nature* in 1995. "The most impressive characteristic of *Giganotosaurus*," he wrote then, "is its enormous size." With a skull about 1.43 metres long, roughly twice that of either *Abelisaurus* or *Allosaurus*, *Giganotosaurus* was the largest known theropod in the world, even bigger than *T. rex*[1] in certain aspects of its upper jaw and some parts of its cervical vertebrae. At the time it was found, its closest relationship seemed to be with the African theropod *Afrovenator*, but subsequent study showed closer affinities to *Carcharodontosaurus*, a theropod from northern Africa, with the best one found recently by the American paleontologist Paul Sereno. Rodolfo placed *Giganotosaurus* among the Tetanurae,

---

1 The largest *T.rex* known is the one they call Sue, now in the Chicago Field Museum. Her bones were the subject of a prolonged custody battle involving the Cheyenne River Sioux Indian Reservation, near which the specimen was found in 1990, the Bureau of Land Management, the FBI, a private Swiss fossil collection company and the Smithsonian Institution. Sue's femur was 1.38 metres long; *Giganotosaurus*'s is 1.43.

one of two large groupings of theropods proposed in 1986 by Yale University's Jacques Gauthier. Gauthier divided all theropods into Tetanurae and Ceratosauria, with the tetanurans comprising all theropods that resembled birds more closely than they resembled ceratosaurians. The tetanurans included such bird-like theropods as *Compsognathus*, *Ornitholestes*, *Deinonychus*, and *Tyrannosaurus*, so placing the huge *Giganotosaurus* in the same group was a bold step. The assignation was based on several aspects of the skeleton, one of which was the tail, which was nearly complete. Tetanurae means "stiff tails," and the group is defined partly by the fact that they seem to represent an evolutionary change from animals that needed long, whip-like tails for balance and locomotion into those with stiffer, shorter tails that, in retrospect, seemed to be pre-adaptations for flight: *Archaeopteryx* had a tetanuran tail, and we can almost see it beginning to bunch up into a pygostyle. The tetanurans had other bird-like features, including air passages in their skulls and furculae in their chests. Since those body parts were missing from the original *Giganotosaurus* skeleton, Rodolfo hoped to find them in this quarry. If this dinosaur had them, then they probably also existed in *Giganotosaurus*.

Although the rocks here and where *Giganotosaurus* had been found were both Late Cretaceous, there may have been a five-million-year gap between them. A lot can happen in five million years. One of the things Phil and Rodolfo wanted to see was how far this dinosaur had evolved from *Giganotosaurus carolinii*. What had changed? Had it become more bird-like? Had the tail become shorter? There was a good chance that the dinosaur itself had become even bigger. A few years ago, Chris McGowan, a paleontologist with the Royal Ontario Museum and one of Phil's former professors at the University of Toronto, played around with some numbers and came up with a surprising theoretical conclusion about warm-blooded creatures: the larger they became, the less they had to eat compared with cold-blooded creatures. A warm-blooded mammal weighing ten ounces, say, needs eight times more energy than a cold-blooded lizard of the same weight. That's why birds, which are warm-blooded, rarely do anything but eat, and lizards never seem to do anything but bask in the sun. But as animals get bigger, the ratio goes down: a one-hundred-pound wolf requires only four times as much food as a one-hundred-pound alligator. If theropods were warm-blooded, then it was to their distinct advantage to get as big as possible, which certainly seemed to have been the case in the Late Cretaceous. It is generally accepted now that the small theropods were warm-blooded; MacGowan's calculations suggest that maybe the really big ones were as well. There must, however, be a threshold at which the size advantage turns against them, a point at which they become so big or so numerous that even their reduced need for food is greater than their habitat can sustain.

We grabbed the picks and shovels and began to work, the writers as well as the scientists. When Rodolfo saw me swinging a pick he seemed surprised, as though he had expected me to just stand around with notebook and cell phone, waiting to report the latest discoveries to the *New York Times*. When he realized I was here to work, he seemed to relax.

"I'm almost sure that this animal was the same size as *Giganotosaurus* or bigger," he said. "I can feel it in my bones. I know that what we so far have is smaller, but ten per cent is well within a normal species range: look at Phil, for example. He's what, six foot four? Knock ten per cent off his height, about eight inches, and he's five foot eight. Lots of humans are five foot eight, even shorter," he said, "and we don't start wondering if they belong to a different species."

"What do we need to find?" I asked him.

He shrugged. "More material," he said, the answer every scientist gives when asked what he or she needs most: more data. "In the museum, when we sorted out all the stuff from last year, it became clear that there are several individuals in this quarry. We have three left tibias, for example, and enough metatarsals for four feet. This is a big bonebed. And yet we still don't have enough to make one complete skeleton. And we don't have a skull."

I told him about Dale Russell's Law of Sauropods, which Dale had formulated in China: the skull is never found. Rodolfo laughed, then he sighed. "Yes, a skull would be nice. Find me a skull. I want this guy's head."

THE ODDS FOR A SKULL weren't that bad, I thought as I shovelled gravel into the wheelbarrow. (It also occurred to me to wonder whether the wheelbarrow had come from Daniel's workshop at the museum, or from one of the exhibits inside; this could easily have been the first wheelbarrow ever used in Plaza Huincul.) Dale's Law of Sauropods didn't apply to theropods. Sauropod heads were ridiculously small for their body size—like a goat's head on a cow—and were poorly attached to their necks. Sauropod brains were so small in relation to their bodies that their heads were little more than entry points for air and food. The skulls of large theropods, on the other hand, were enormous, much bigger than even their huge bodies would suggest was necessary, and for some reason, their skulls seemed to have remained attached to their bodies after death. If there were four or more individuals in this quarry, there was a good chance we would find a skull.

After an hour or so of scrabbling at the sandstone with picks and shovels, it became clear to us that the wheelbarrow, even if it held up under the indignity of actually being wheeled with a load of gravel in it, was inadequate for our needs. One wheelbarrow, even a brand new one, could not have kept up with the amount of gravel produced by three picks and three shovels. And ours was not new. It wasn't even merely old. It was decrepit. It was a cruelty to ask it to perform at all. Each time Christian struggled off with it towards the edge of the hill, we would watch it go with its oddly oblong wheel, its jangly box, its creaking arms, and wonder if there was really any point in his bringing it back. We needed another system. I don't know who came up with one. It might have been Val, who wanted something to do, or it might have been the ever-practical Mike, or it might have been Daniel, unable any longer to witness the abuse we were heaping on the wheelbarrow, but someone remembered that there were half a dozen plastic wash basins down in the kitchen tent, brought along for such obscure domestic purposes as washing dishes and tossing salads. These basins, it was suggested, would be perfect for hauling eighty tons of gravel. Six people with six plastic basins could easily keep three shovellers busy, and we could still continue using the wheelbarrow until it finally wobbled off by itself behind a hill to die.

This is how we managed it. Three of us would bash at the sandstone slope with picks until our arms were too weak to raise above our heads. Then we would bash at them some more from shoulder height. Then from the waist. This would take ten or fifteen minutes, and would dislodge a pitifully small amount of gravel, a few piddly bucketfuls, from the conglomerated sandstone. Then, while the three pickers collapsed in a heap on the ground and three more took their places, one person with a shovel would scrape the accumulating rubble off to one side of the quarry, where two others with shovels would heap it into the plastic basins set on the ground for the purpose. Each basin held about three shovelfuls of gravel. When a basin was full, someone would pick it up, carry it to the edge of the quarry, tip it over the side, and then replace it at the refilling station. Meanwhile, the pickers, too exhausted from their efforts to resume picking, would have changed places with either the shovellers, who were only slightly less exhausted, or with the chisellers. These wielded the two-and-a-half-pound hammers,

driving cold chisels as deep as was feasible into the solid conglomerate, pounding the chisels sideways to loosen them in the matrix,
and then withdrawing them so that the pickers could come along
and finish the job. It took about a dozen strokes with the hammer
to drive the chisel three inches into the rock. The fact that these
methods would not have seemed advanced to paleontologists
working in the Montana and Alberta Badlands a hundred years ago
did not escape our notice. Nor was it lost on us that convicts in
Kingston Penitentiary, breaking rock for city roadbeds seventy
years ago, had used better tools and more effective methods than
ours. It was brutal and backbreaking labour, but it felt wonderful.
We loved it. We laughed all morning. We smiled as we passed one
another with our red and yellow plastic basins. We shared our
water bottles without first wiping the spouts. There was not a harsh
word or a complaint spoken. As the morning progressed and the
sun rose above the cliff and the hilltop moved almost impercepti-
bly down, the work warmed us even more than the sun did. Jackets came off, then sweaters, then overshirts. Sunscreen and
wide-brimmed hats came out. By twelve o'clock, when Daniel
called the morning maté break, the thermometer in the quarry
read 30 degrees Celsius.

DANIEL WAS OUR *cebador*, the one who performed the pouring of
the maté, a very involved process, not unlike a Japanese tea ceremony in its ritualized intensity, that I watched with fascination. He
filled the maté gourd with yerba and soaked the herbs in cold water
while the kettle heated. Then he sucked the cold brew through the
bombilla and spat it over the side of the hill. The first taste of maté is
always bitter and filled with yerba dust, and so the *cebador* must deal
with it himself. Daniel repeated the ceremonial spitting several times,
as is the custom in Patagonia, until he was satisfied that the maté was
sufficiently steeped, then refilled the gourd with hot water and passed

it to Rodolfo, who sipped meditatively on the *bombilla* until the water was gone. Then he handed the gourd back to Daniel for refilling. Daniel poured more hot water into it and passed it to Phil.

The rest of us lay back on the hillside and waited in the sun for our turns. We were almost too exhilarated from exhaustion to talk. I was reclining next to Don Lessem. As well as being a writer,[2] Don was one of the founders of the Dinosaur Society, a nonprofit organization designed to increase public awareness of, and thereby raise private money for, paleontological expeditions like this one. Don had helped set up the Society in 1991, and then retired from it a year later. For a while it continued running, but by 1996 it had more or less settled into diapause. I was curious about what had happened.

"Well," Don said, "it sort of collapsed in on itself. The society raised a lot of money for research right off the bat. We got the sets and props from *Jurassic Park* from Universal Studios and Steven Spielberg, along with casts of the dinosaurs made for us by Peter May, the whole bit, and we put them on a tour of fifteen cities across the United States, from New York to San Diego. We raised $3 million in the first year. But only a third of that ever got distributed to scientists. The rest went to overhead. Obviously, something wasn't working."

A few years ago, Dale Russell had estimated that, worldwide, a total of only $1 million a year was spent on paleontological field research from all sources. This meant funding for only a handful of scientists,[3] and with government cutbacks, fat-trimmings, program

---

2 His book, *Kings of Creation*, came out in the States a few months before my *The Dinosaur Project* appeared in Canada, and covered pretty much the same territory. His looked at how paleontology in general was changing as newer and more exciting dinosaurs were being found around the world. Mine focused on the Canada-China Dinosaur Project, and examined the ways in which paleontology was changing as more paleontologists with backgrounds in biology and ecology, rather than strictly geology, came into the field. Don had visited Phil's camp in Inner Mongolia in 1988.

3 Dale estimated that the $1 million supported about one hundred scientists, only thirty-five of whom got out to do actual field work. Since there were fifteen of us at the Plaza Huincul site, we therefore represented nearly one half of all those who would be digging for dinosaurs that year if it weren't for private organizations like the Dinosaur Society. This is an observation worth keeping in mind.

chops and priority changes, things had gone downhill since then. The money spent on research by the Dinosaur Society doubled the world figure for that year, which I thought was not bad.

"True," said Don, "but it was only for that year: there was no new revenue coming in. After those fifteen cities, that was it."

By the time Spielberg's second dino-blockbuster, *The Lost World*, came out, Don was looking for greener pastures. He signed a deal with Spielberg and Universal to organize another travelling dinosaur show that was twice as big and three times as expensive as the first one. To handle the business, he and Val set up a new organization called the Jurassic Foundation. ("Universal wanted it called the Jurassic Park Foundation," he said, "but Spielberg himself nixed that.") The foundation designed an exhibit that did more than simply ship around casts of Spielberg's model dinosaurs; it also explained the science behind the two movies, the steak beneath the sizzle, everything from chaos theory to DNA sequencing. When Spielberg's giant *Velociraptors* breathed visible breath in *Jurassic Park*, for example, that was his way of illustrating the theory, outlined in the exhibit, that small theropods were probably warm-blooded. The film also showed the fleet, herbivorous *Hypsilophidons* complacently munching (Cretaceous) grass in huge herds, at least until a rogue *T. rex* disrupted the pastoral scene; that, too, was Crichton's way of saying he had accepted very recent evidence that some herbivores gathered in herds, laid their eggs in vast breeding colonies, migrated to summer feeding grounds, and generally behaved like modern gazelles. For $300,000, Don explained, a museum could bring in a 12,000-square-foot exhibit that explained the complicated engineering behind the film's paper moons, and keep it for three months. The exhibit went to New York, Cleveland, Houston, Baltimore, Columbus and Mobile, Alabama. The museums made money from it, Don said, and the Foundation was able to provide $55,000 for research in 1998. I did some quick math and anticipated Don's next statement: "We're more modest than the Dinosaur Society was," he admitted, "but it looks like it's working."

The money, he said, was distributed by a small jury of scientists who voted on grant applications—this year the jury was Phil, Jack Horner and a New York paleontologist, Cathy Forster. Eva and Val took care of the administration, and Jack Horner's wife, Celeste, set up and maintained the Web site. Phil was the scientific head of the Foundation, the name that gave it its credibility. "Jack Horner was perhaps the more obvious choice," Don said candidly, "since he was the chief scientific adviser for both movies. But Jack has a reputation for having a blunt manner, and he has little patience for administration. We needed someone with more skills in diplomacy. Phil gets along with everyone. Rodolfo had offers from dozens of paleontologists to come down here to work with him. This is the hottest place on the planet to work. But he is, how shall I say it, very sensitive to American scientific imperialism. Other scientists come down here, throw money around, think they can run the show, then take the material back to the States with them. Rodolfo ends up being second author to whatever paper eventually comes out on work done on his own turf. He knew that wouldn't happen with Phil."

I didn't know what it was, but something nagged at me as we went back to work in the quarry. It was as though a minor chord had been struck in the sweet symphony of science I had been listening to so far. The feeling kept badgering me as I shovelled more gravel into a veritable *Fantasia* of plastic washbasins. It had always been my view that the proper source of funding for scientific endeavours was the government: governments gave money to their public institutions, like the Tyrrell Museum, and the institutions distributed it to worthwhile projects, such as Phil's trip to Argentina to study theropods with Rodolfo. Phil's work ultimately benefited society, and so it made sense to me that society should pay for it. This of course was a simplified, possibly even naive, version of how things worked, but it represented to me the broader outlines of how pure science (and for that matter art) ought to get done in an enlightened, liberal-minded society. The social benefit of pure science to everyday life may not be as direct or obvious as, say, hydro dams or Medicare, but then the social benefit of lots of things governments fund isn't always blindingly obvious to me. Putting long-range assault weapons on space stations, for example. But without pure science, no hydro dams would be built and not much medicine could be practised. (And there would certainly be no space stations.) Don had brought home to me that we do not live in an enlightened, liberal-minded society. Governments do not give enough money to their institutions, and scientists have to look elsewhere for funding if they want to continue their researches, to private corporations, for example, or individual philanthropists. This seemed to me a particularly dicey way to finance a long-term scientific project. Corporations, in my experience, are not known for their enlightened, liberal-minded attitudes towards the disbursement of money. Corporations, and even individual philanthropists, have their own agendas: corporations want their logos emblazoned on smiling skiers' foreheads ("I know, let's call it the Jurassic *Park* Foundation!"), and even anonymous philanthropists want tax write-offs. When a skier stops spending a large proportion of his or her time before an admiring television camera, or when the noose tightens around a tax loophole, corporate and philanthropic money tend to drift away in search of more congenial hosts. Governments are always, in an ideal world, other-directed; corporations, especially in a world that is less than ideal, are always self-directed.

In fact, I had only to think of the premise behind *Jurassic Park* to appreciate the irony in the situation Don had described with such enthusiasm. In the film, the wealthy head of InGen Corporation,

John Hammond, approaches a cash-strapped paleontologist, Alan Grant, and offers to fund his field work in Montana for the next five years if Grant will sell his reputation for scientific integrity by endorsing InGen's money-making enterprise on Isla Nublar. This is indisputably a Faustian bargain: Grant is being asked to sell his soul to the devil in exchange for earthly riches, never mind that he intends to use those riches to further the cause of pure science. After striking such a bargain, his soul is no longer his own and the science can never be pure.

WE WORKED FOR TWO more hours before anyone even mentioned lunch. Finally, at two o'clock, Daniel set down his pick and said something to Mike, and the two made their way down the hill towards the kitchen tent. The rest of us kept working. Those who had been picking and shovelling had blisters on their palms, and blisters inside blisters. The fingers of those who'd spent more time hauling gravel in the plastic washbasins were torn and bent from grasping the lip around the basins. I had taken shifts doing both, so I had blisters and ragged fingertips. The work gloves were in the famous crate that was still sitting in Customs in Neuquén. It was surprising how cruel a washbasin could be. Some, we discovered, were crueller than others. The red ones dug into our fingers more than the green ones, but the green ones didn't hold as much gravel, which meant more trips. The yellow ones had larger lips, but held even more than the red ones, which meant heavier loads. It became a macho thing to pick up a yellow washbasin without wincing, and carry it to the edge of the quarry, dump it over the side in one fluid, graceful, athletic motion, like a Scotsman tossing a caber, and to turn back for another load without pausing to watch the splay of fresh gravel fan out down the hillside. Such a pause, though it answered a purely aesthetic desire to admire the results of our work, might be interpreted as needing a rest.

At three o'clock, Mike called up from below that lunch was ready, and we all relinquished our various labours and filed down to eat. Mike and Daniel had made a pot of soup by boiling some vegetables over a fire and adding several packages of freeze-dried soup mix to the water. There were dozens of these packages in the supply box, no two alike, so the soup was an interesting mélange of rice, noodles, chunky chicken, onions, carrots, potatoes, beef broth and tomato pasta. The table was laid out with slices of bread, cheese and a kind of sausage. There were also large bottles of Quilmes beer and two rectangular boxes, called bricks, of Argentine wine, one red and one white. We ate and drank with mucho gusto. We had done a good morning's work.

Rodolfo and I stood at the edge of the canyon, munching sandwiches and drinking soup out of our tin cups, looking up at the quarry with contemplative satisfaction. The top of the hill was discernably lower than it had been at breakfast; the basins of gravel we had spilled over the side formed a miniature scree deposit, a buff-coloured apron that fanned out over the red hill from the lip of the quarry to about halfway down the slope. Rodolfo looked up at the deep blue sky, at the smattering of thin white clouds; they were moving slowly over our heads from west to east.

"How much longer until we get down to bone level?" I asked him, just for something to say.

"Not long now," he said, looking back at the hill. "Two, maybe three days."

"Days?" I said, startled. I had been thinking a few more hours.

"Where we've been digging so far," he said, "the congolmerate has been very loose, we just have to hit it to break it up. The work has gone very fast. But where the bones are, where they disappear into the side of the cliff, there the rock is very hard." He shook his head. "It looks as though we're digging in a channel deposit, with the centre washed out of it. The bones are around that centre.

Maybe they were lying on the shores of an ancient river. Anyway, the centre is soft, filled in with gravel that has become hard but is not yet solid rock. But around the edge, the ancient shore, it is sandstone. Very hard," he said, taking a drink of soup and looking dubiously into his cup. "If we had a jackhammer," he said, "we could do it in a few hours. But all we have are our bare hands and these picks and shovels. And I hate shovelling," he said vehemently. "This afternoon, I'm going to work on the bones that we left exposed last year. I'll leave the shovelling for you younger guys."

I pointed out that I was ten years older than him.

"You know," he said, turning to me. "I hate the idea that I will be forty next month. I feel like an old man, like my life is over."

"Thanks a lot."

"No, really, maybe it's not the same for you. But for an Argentine, to turn forty is to begin to die."

I told him that for me, turning fifty had not been attended by any such existential angst. My wife had thrown me a big party; she gave me a mountain bike and my friends gave me bottles of Scotch.

Rodolfo regarded me with a look of profound pity. I felt he was warming to me. "I told Claudia that we will not be celebrating my fortieth birthday. But I know I am becoming an old man. For example, I find myself developing an interest in tango." He said this in the mournful tone an Arctic explorer might have used to report that one of his toes was dropping off from gangrene. "In Buenos Aires, no one becomes interested in tango until they are too old to do anything else. You know? If you can't have sex for real, you dance about it. Not tourist tango," he said. "Tourist tango is only about skill, all that twirling and throwing each other around. Real tango," he brought his fingers to his lips, "real tango is poetry, a kind of secret language. They had their own words for things, for police, for prison, for sex, so that no one from the outside knew what they were talking about. But tango culture is dying now. The last true artist of tango was Piazolo, in the 1950s. No one is writing tango any more."

"Tango is going the way of the dinosaurs," I said.

"Right," said Rodolfo. "And I am going the way of tango."

AFTER LUNCH WE climbed back up to the quarry and resumed our delving. Rodolfo worked on a bone that had been discovered last year and left partially buried to provide a benchmark for this year's dig: a huge tibia, nearly a yard of it visible near the lip of the bonebed. Still partly embedded in rock, it looked like a gigantic dog biscuit sticking out of a pile of sand. Rodolfo selected an awl and a geology hammer from the box and worked at the end that was encased in matrix. The rest of us continued to bash away at the hilltop. Paul and J.-P., back from their geological ramble, were now helping out in the quarry; meanwhile Phil, Paul and I were at the very top of the hill, where it met the cliff, carving out a narrow, level path for the wheelbarrow. The wind had abated, the sun beat hotly down upon us, and our view across the valley to the opposite rise was stupendous. High yellow cliffs rose above a level floodplain sprinkled with thornbush and pale green and brown *coirón* grass. A good place to look for birds, I thought. Above the plain, a group of white figures were picking their way across a cliff face.

"Ghosts," I thought Paul said.

"Ghosts?" I asked.

"Goats," said Phil. "The farmer lives just up the valley a ways. We went to visit him last year. He gave us maté. Lives out here by himself all year round. He'll probably drop by one of these days to see who we are and what we're up to."

As I watched, more goats appeared from around a bluff, and soon there were at least a hundred of them threading up the valley wall opposite the quarry. What on earth did they survive on? But beneath the thornbushes I now noticed another type of vegetation, softer and more grass-like than the spikey *coirón*. It was invariably cropped short. The goats must live on that.

"They pass by the quarry every few days," Phil said, "coming over the cliff here and down to the river bottom, then up the other side. They seem to have a regular route they follow."

We went back to work, and the late afternoon evolved into early evening without our being conscious of it. My body had long ago stopped being aware of anything as abstract as the passage of time. But at precisely seven o'clock, a crimson glow emanated from somewhere below the hilltop and spread like heat across the sky. Never had I seen such a sunset. The whole western portion of the sky turned a vivid red-orange. Then darkness set in rapidly, reminding us that we were, after all, in the subtropics, although the temperature fell so rapidly with it that tropical was not the word that sprang to mind. By the time we had put the tools away, put our warm clothes back on and set off down the path to the canyon, it was too dark and cold and late to look for birds.

CHAPTER SEVEN

Beyond the Dusty Universe

I T WENT ON FOR days, this digging and delving. Our hands toughened, our backs stiffened, our skin darkened. The quarry began to take on a definite shape. There was still a nugget of very hard concretionary rock in one corner that we all surreptitiously avoided. Phil said he didn't think the bonebed extended that far anyway, but of course we all knew it would and that sooner or later we would have to tackle it.

Meanwhile, we dug. Rather than continue skimming off the top of the quarry to get down to bone level, as the rest of us were doing, Daniel chose a spot at the edge of the quarry, almost at the lip of the hill, and dug straight down. After half a day he was waist deep in a narrow pit about six feet long and two feet wide, well below gravel level, pulling out huge blocks of sandstone and piling them up beside the hole. No one said anything, but references to "Daniel's tomb" began to creep into conversations. "Where's the shovel?" "Over there by Daniel's tomb." When he was down to chest depth, Mike finally broke down and asked him in Spanish what his plan was. Daniel shrugged. He seemed merely to have gotten himself into a rut. Mike told him that if he dug just a little deeper and then began to tunnel in towards the quarry, he might come up under the rest of us and get at the bones from underneath. Daniel looked thoughtfully into his hole, shrugged again, and kept digging.

One day, Rodolfo took the truck into Plaza Huincul to deposit Don Lessem and Val at the bus station and to pick up more supplies. We had run out of our three staples: water, bread and beer. He was also going to check on the status of the box of implements that had been impounded by Customs.

When he came back at breakfast time the next morning, he had the boxes. "Seven hundred pesos!" he shouted up from the truck bed. Phil and Eva exchanged grimaces. Rodolfo also brought three new crew members: Jørn Hurum, his wife, Merethe, and her sister Inge. They were from Norway, although for the past year Inge had been living in Chile, where she and her husband ran a fish farm. They had just spent nineteen hours on a bus from Santiago and looked thoroughly bedraggled. Jørn, a paleontologist finishing his post-doctorate, under Phil's supervision, at the University of Oslo, was a young, jovial, likeable fellow who instantly endeared himself to us by producing his contribution to our reduced larder: two bottles of Argentine whisky, of a brand called Breeders' Choice. "I got it for its label," he explained, pointing to a banner under a realistic portrait of a black Angus bull: "'Made by breeders, for breeders,'" it read. "It's very cheap and not bad, if you mix it with coffee and hot chocolate."

The next day, it began to rain again. This time it wasn't a warm, life-giving, desert rain, but a cold, wind-driven, winter rain. During the night the weather must have shifted to the east. I had become used to sleeping with the tent flapping like a tethered banshee around me, but when I woke this time there was a steady, staccato, snare-drum roll in the darkness. When I realized it was rain I groaned, not for the first time that week, and tried to go back to sleep, but was kept awake by the syncopated din. There was also a faint gurgling sound that seemed to be coming from somewhere beneath the tent: water was already accumulating in the small canyon. This realization also made sleeping difficult. I felt for my flashlight, turned it on, and picked up a book. It was George Gaylord Simpson's *Attending Marvels*. With the flashlight tucked under my chin, I opened the book at random and read this cheering little paragraph: "As we came home the sun went out suddenly and the whole land turned a sinister gray, dark and light but without a spot of color. Streaked vicious clouds poured over us like a flood from the west. The moon rose yellow through the last band of clear sky. Rain began to patter, then to pour." This must have been what it was like last night, after I went to bed, except in this part of Patagonia the foul weather came from the east. The sky the previous night had been exceptionally clear, I recalled, beginning with the six pale stars of the Southern Cross in the evening and, as the night progressed, moving to Orion standing upside down and a crescent moon with the crescent on the wrong side. Curiously, Simpson's thoughts mirrored my own: "Beyond this element-tormented spot," he continued, "lies vast, desolate Patagonia. Beyond Patagonia lies the world of seas and plains and mountains, for complacent thousands of miles. Beyond the world wheels the dusty universe."

Breakfast was a bleak affair. We all crammed into the kitchen tent, which roared and pounded in the furious wind, and sat on damp cardboard boxes with bowls and cups balanced on our knees. Outside, the rain came down in torrents, so that Daniel had trouble getting a fire going to heat water for coffee and maté. The only consolation was that someone had left the radio outside all night, and now it didn't work. I took a perverse glee from this. Daniel was taking it apart to let it dry, but he was shaking his head funereally at its loss. It belonged, it seemed, to his aunt, who prized it above all else, including her nephew. Phil and Eva were conferring silently in one corner while the rest of us sat around the rickety plastic table loaded with the by-now familiar breakfast goodies. Out of sheer boredom, Jørn and I watched J.-P. making his breakfast. He was limping badly from a recurring case of gout and wearing a pair of woollen gloves with the fingers cut off, and both the limp and the gloves lent his preparations a furtive, Dickensian air. First he stirred a bowl of water and instant oatmeal into a thick, grey sludge, then, discovering we were out of milk, added a few tablespoons of *dolce de leche*. Pulling the spoon out with some difficulty, he then tasted the mixture. "Something miss-ing," he muttered. Jørn took the spoon and tasted it.

"Tastes like wallpaper glue," he pronounced.

J.-P. agreed. "Too dry," he said. He added a heaping tablespoon of Nestlé's Quick to it.

"That's powder," said Jørn.

"You're right. Still too dry." J.-P. looked sadly into his bowl for a second, not liking what he had to do but knowing that he had to do it. Then he poured his coffee into the bowl and stirred it around.

"That even *looks* disgusting," said Jørn.

J.-P. licked a protruding finger and nodded. "Not bad," he said.

For several days now he and Paul had been going out in the mornings, pacing along the bases of cliffs with tape measures, climbing up rock walls, clawing at screes with J.-P.'s Bunyan-sized pick, and generally looking at sharply rising mounds of sand and gravel that had been dumped there eighty million years before, try-ing to figure out what dumped it and why. The whole valley was sprinkled with their markers. I asked J.-P. what conclusions they had come to and what they intended to think more about later.

"Well," he said, "the rock in the quarry is definitely polymictic, matrix-supported conglomerate."

"That's hard to say with a mouthful of oatmeal. What does it mean?"

"Oh, uh, let me see." He taught geology at the University of Cal-gary, so I figured he could put it in lay language for me. "Poorly sorted clasts—pebbles, that is—of numerous types and sizes, but all of them sort of floating in coarse sand, touching sand more often than each other."

"Which means what?" I said.

"Oh, lots of things," he said, nodding. I could see he was wondering how far back he'd have to go with this before he hit something I understood. "There's trough cross-stratification: we're looking at a river channel here. Suspension deposition, numerous lateral accretion surfaces, which is why you see parallel lines in the sandstone. That means the bar was moving out."

"So the bones were washed in from somewhere else?"

"Yup. Uh-huh."

"From where?"

"Well, upstream would be my guess."

Before breakfast, Daniel, Christian and Adrian had pitched a tarpaulin, a large, green, plastic affair measuring ten metres square, over the quarry to protect the bones—and incidentally us—from the rain. It was rather ingenious of them. By now the quarry was sufficiently below ground level that one end of the tarp was simply laid out at the top of the cliff and held in place by large rocks. The opposite side was affixed by guywires to more rocks and several convenient thorn bushes, and held up at the corners by poles made from short pieces of lumber wired together, with a rope running down the middle to form a peak. Rainwater ran down from this peak and collected in pools on the tarp. We tried to save some of it for washing; we were, after all, in a desert.

Some of us were still picking away at the rock along the cliff wall, while others had taken up finer tools and were carefully easing our way down to where we suspected the bones were. There were thus three or four activity centres in the quarry, foreshortened by the lines of the tarp and the variously coloured layers of rock and gravel, as though we were working in an Escher drawing. Pale daylight filtered through the green plastic, lending the quarry an eerie luminescence, like an operating theatre under mercury lamps, and the sound of the rain rattling above our heads was so amplified that even a minor sprinkle sounded like a major deluge. But at least it drowned out the radio, which was working perfectly again. Daniel's aunt would be pleased.

Phil, wearing a yellow plastic rainsuit, was stretched out on the quarry floor, picking at something with his awl. "Looks like a haemal spine," he said when I asked him, leaning back to show me a length of curved, blade-like bone lying flat on the surface. "The part of a tail vertebra that extends downwards." He was carefully clearing away the rock from around the spine, and asked me to extend the cleared surface out farther, being careful to check for more parts of the vertebra. With some trepidation, I began scratching at the quarry floor beside him with an awl and sweeping away the shards with a paint brush. Now that we were down to bone level, we'd set the sledgehammers, picks and chisels aside and taken up more delicate tools. We removed the rock a flake at a time. First we swept a small area with a two-inch paint brush and examined the surface for bone. If there was no bone, we scored the surface again with an awl; if there was bone, we switched to a dental pick and a one-inch paint brush. We weren't looking for shape so much as for texture, a peculiar shade of stone, a certain bone-like intrusion of the senses. Fossil bone comes in a wide range of colours, rarely bone white, more often the same colour as the matrix that encases it. Brushed sandstone looks a lot like bone, and sometimes the only way to tell the difference is to break off a chunk and hold it up to the eye. If the broken edge looks like packed sand, it's sandstone; if it's finer grained than that, it's mudstone or siltstone; if, however, it's honey-combed like marrow, if, in your imagination, you can see blood oozing through chambers, then it's probably well not to have broken off too big a piece. Your eyes are constantly straining to see anomalies in stone, subtle colour changes, a smoothing out of texture, a sudden whorled pattern in the depositional chaos. Meanwhile, your tools are ploughing through the matrix, brushing away chips and shards, pulling out stones and roots. I learned to hate roots. Roots follow natural fissures in the rock, sinuating down through the ancient soil to find water, then wicking water back up through the rock to nourish the long-disappeared plant they belonged to. Minerals from the water seep through the root walls and harden on their rounded surfaces, making them look exactly like bone. Again and again, I was fooled. My heart would stop, my adrenalin would rush, I would murmur something like, "Hmmm," which would make Phil look up to see what I'd found, and then he would say that what I had found was only a root. Worse than that, roots often extend down through the rock by growing inside the marrow of buried bones. Sometimes, after slow and painful scraping with a dental pick, I would discover not only that I had been excavating a root, but that I had broken through real bone to get to it.

After days of measuring our progress in yards, in feet, in inches, we were now lowering the quarry floor a fraction of an inch at a time. It took an hour to fill a plastic washbasin, and then, for a break, we'd get up and dump it ourselves, cursing our joints, slipping on the wet hillside beyond the shelter of the tarpaulin, then stumbling back to the quarry. We worked on our knees, propped on our elbows, lying on our sides, getting as close to the damp, cold rock as we could. It was like digging for land mines. At any moment a bone could burst up through the ground and grab your heart. The bone was locked up in the rock, and we were listening for the tumblers to fall.

THE SCIENCE OF WHAT happens to a body after death is called taphonomy. Think of a dinosaur dig as a murder mystery; the quarry where the body is found is the scene of crime, and taphonomists are forensic detectives trying to figure out if the victim died there or was transported, and if the latter, where it was transported from, and by what vehicle. A few years ago, John McPhee wrote a story for the *New Yorker* about forensic geology. By checking such clues as pebbles caught in the treads of a tire or grains of sand in a victim's pant cuffs, and comparing these against known geological deposits, geologists could sometimes tell the police where a car containing a victim had been before it got to where it had been found, and what route it had taken to get there. This is a form of taphonomy. Like paleontology itself, taphonomy is a fairly imprecise discipline, a soft science, depending as it does on the subjective interpretation of usually incomplete evidence. Looking at a layer of sandstone and trying to figure out how it got there is stratigraphy, and even stratigraphers refer to it as voodoo geology. Paul and J.-P. were forensic geologists trying to shed light on a death that had taken place more than eighty million years ago. The scent was cold. But the evidence was still in place, the possible means of transport finite, and a tentative explanation within reach.

The trick was not to go backwards in time, but to think of the past as the future. We could not recall; we had to imagine. Somewhere close to this spot, somewhere upstream, half a dozen extremely large theropods were hunting or scavenging—possibly they'd picked up the scent of a dead sauropod like the one buried under this very quarry—when they were caught in a sudden torrential rainstorm. Maybe they were walking along the edge of a canyon when the wall collapsed and they all fell in. Maybe they were fording a large river and were hit by a wall of water. However it happened, they all died at the same time and their bodies, more or less intact, were carried downstream by the swollen current until the river, slowing down to round a curve, deposited them here at the quarry. The carcasses lay exposed to the elements for a while, long enough for their soft parts to rot away and their bones to be picked clean by insects, pterosaurs, birds and other scavengers, and their skeletons to be bleached and weathered, and then slowly the bones were covered by sand.

"Why would half a dozen gigantic theropods be travelling in a pack?" Phil wondered aloud.

We were having our morning maté break, this time under cover. Fortunately, Daniel didn't mind making maté in the rain. He'd found a level ledge partway down the hillside, sheltered from the wind by a huge thorn bush. The fire burned merrily in its ring of stones, the blackened kettle nestled upon it. I was beginning to suspect there were no conditions under which Daniel would mind making maté.

I thought about how much meat a pack of carnivores the size of *Giganotosaurus* would need to sustain itself. A lot, I decided. If a one-hundred-pound wolf wolfs down ten pounds of meat a day, then an eight-thousand-pound, warm-blooded dinosaur could require up to eight hundred pounds. Half a dozen maurauding reptiles, if they were warm-blooded, would have to bring down a sauropod every week to ten days, which was about as often as a pack of wolves kill a moose.

There was no doubt we were dealing with a pack. This was not a random assemblage of bones. In a normal bonebed, a deposit at the mouth of a river, for example, where bones collected like driftwood from far inland over many years, theropod bones make up five to ten per cent of the total. Here it was one hundred per cent, and all the same species, all in the same state of disarticulation. "Clearly very interesting," Phil mused, in his best Holmesian manner. This did not, however, appear to be a three-pipe problem. "Obviously a single incident, one event that killed them all. Fire, disease, poison, trampling by stampeding sauropods." He didn't rhyme these off very convincingly. "Could have been drought," he conceded. "Paul and J.-P. say they feel it was reasonably dry around here. Maybe this was a watering hole, and the pack just hung around waiting for thirsty sauropods to happen by, and then they'd pounce. But then we'd expect to find sauropod bones mixed in with the theropods, and we don't. No, I'd say it would have to be something like a flash flood."

By the time we left the quarry that night the rain had coated the hillside with wet Plasticine. We skiied down it to our tents, our boots so caked in red ooze that by the time we reached the bottom we could no longer see them. The danger of course was that we would slide down so fast we'd be unable to prevent ourselves from plunging into the water-filled canyon at the bottom, but somehow we avoided this. We didn't know it then, but it would continue to rain for the next week, a steady, pelting drizzle that made it impossible to leave the camp for even a walk, and turned the short path to the kitchen tent into a quagmire of boot-sucking, slippery goop that threatened with every step to propel us down its slight incline and over the edge of the gully into the raging torrent. We took our cold, wet clothing off at night, climbed into our cold, wet sleeping bags, and put our cold, wet clothing back on in the morning. What was the point of dry socks when to put on a boot was to shove your foot into a shapeless blob of mud-soaked leather? The morning temperature had risen slightly—to four degrees Celsius—but I discovered that a sleeping bag rated for minus twenty was warm only when it was dry. I abandoned George Gaylord Simpson and took up Gerald Durrell's *The Drunken Forest*, leaving the wind-blasted, rain-soaked Patagonian wasteland behind to search for birds in the warm, subtropical sunlight of the Argentine rainforest, well to the north of us.

"HERE, TRY THIS." Jørn handed me a cup filled with a taupe-coloured concoction. Steam rose from it, so I drank it. It was good.

"What is it?"

"Cocoa, coffee, *dolce de leche* and Breeders' Choice."

I wrapped my hands around the cup to warm them and sat down on a slightly squashed cardboard box, one of those that had come out of the crate recently rescued from Customs. It seemed to contain a dozen rolls of wet paper towels. It was the third straight night of rain.

"Don't sit on that," said Mike.

"What is it?"

"A dozen rolls of wet paper towels."

I moved to a canvas and aluminum camp chair that was normally very comfortable (a going-away gift from my daughter) but which I now discovered was situated directly under a hole in the tent roof. The kitchen tent had sprung leaks everywhere. The wind had ripped some of the guy-ropes out of their sockets, and water pooled on the roof and dripped through onto the chairs, the boxes, our heads. Water coursed along the mud-covered floor; looking down at it reminded me of flying over the Mackenzie River delta. A Coleman lantern, suspended from one of the cross-poles above our heads by a short length of wire, rocked gently, dripping water onto the table. The rain outside was like liquid wind. I sat on my wet chair and drank Jørn's elixir, admiring the way the swinging Coleman lantern made the shadows on the table, pinned at one end by the objects they were shadows of, rotate back and forth like windshield wipers.

Merethe was sitting beside Jørn, talking to Inge in Norwegian. In the quarry, she and Eva had been conversing all afternoon, Merethe speaking Norwegian and Eva speaking Danish, without the slightest difficulty understanding each other. Jørn and Merethe and Inge had brought ingenious little Scandinavian camping kits, cup, bowl, pot and utensils all folding up and fitting inside one another to make a compact unit not much bigger than a sardine can. Her cup, from which she, too, was drinking Jørn's whisky-laced brew, collapsed when not in use to a flat plastic disc that no doubt played ABBA songs when slipped into a solar-powered travelling IKEA compact disc player that doubled as a pop-up toaster. I was drinking from the same cup I had had in China, a huge white-and-blue porcelain job, sadly chipped but to which I was fondly attached. It had an almost limitless variety of uses, coffee being one of the least imaginative of them. I had shaved from it, brushed my teeth with it, washed my hair with it, scooped sand out of the quarry with it. With three of these packed with some of J.-P.'s oatmeal and a bit of rope, I suspect I could have made a serviceable *boleadora* with which to bring down a rhea, if I ever encountered a rhea. In *The Drunken Forest*, Durrell describes a rhea hunt, probably one of the last, with gauchos on horseback swinging *boleadoras* over their heads and launching them at the flee-ing birds so that the rope became entangled in the birds' legs and around their necks. Maybe I could practise on J.-P. and Paul the next time they went off geologizing.

I was having no luck in the rhea department. Several times, before the rain, I had spent my lunch hours wandering around on the flatlands above the quarry, picking my way through the thorns with binoculars and bird book in hand, hoping to scare up a rhea or two, but except for an occasional shard of eggshell, I saw no sign of the creature. As Durrell had remarked, they were smart birds, and they had learned to be wary of humans. The male "looked like a small, grey ostrich, with black markings on his face and throat. But his neck and head were not bald and ugly, like the ostrich's, but neatly feathered; his eyes had not the oafish expression of the ostrich, but were large, liquid, and intelligent." Durrell had puzzled me in the matter of toes, however. When his gauchos succeeded in lassoing a rhea in one of their *boleadoras*, Durrell went up for a close look. He noted the "great muscular thighs, like a ballerina's," which I have to take on faith, never having, alas, seen a ballerina's thighs from close up. "The wing-bones," he said, "were fragile and soft, for they could be bent like a green twig." No mention of claws. He does, however, say that "the large feet with their four toes were thick and powerful." By "four toes," he must have meant the three forward-pointing toes and the one reversed toe, which in ratites is so reduced as to be virtually nonexistent, registering more like a heel than a toe. The matter of toes, how many there were and which ones were missing, is important to the bird-dinosaur debate.

All the early dinosaurs had five toes, five being the most common number of digits in Earth's very limited variety of body plans. As theropods evolved, the two outside toes (digits one and five, big toe and baby toe) dwindled in size and importance, until eventually the big toe moved around to the back of the foot (as in apes, and for much the same reason: so the animal could grasp things with it), and the pinky became a sort of left-over appendage way up at the ankle, where it couldn't have been of much use for anything but grooming. Finally, in birds, it disappeared altogether, and the big toe, now completely reversed, could be used for balance while walking, for stability while perching on trees, and for efficiency while grasping prey (think of an eagle flying off with a salmon in its talons). The tarsals, the long bones that join the toes to the ankle, fused into one bone. As early as the 1860s, German embryologist Carl Gegenbaur studied the development of a bird embryo from conception to hatching, and noticed that, as the embryo matured, its foot went through exactly the same stages of development as the dinosaur foot did over millions of years.[1] At first, the bird embryo's foot has five toes. By the time the bird hatches, digit one is reversed, digit five is gone, and the tarsals and metatarsals have fused. This provided T. H. Huxley with some of the ammunition he needed for his claim that birds derived from dinosaurs.

The rest of Durrell's description of the captured rhea might have been describing a feathered dinosaur, perhaps an escapee from

---

[1] The scientific expression for this is, ontogeny recapitulates phylogeny: the embryo goes through in a few months all the stages of evolution the species has experienced since the beginning of life. After Gegenbaur, another German, Ernst Haeckel, examined human fœtuses and recorded the same phenomenon: we are first fish, then tadpoles, then primitive mammals, and only become human a few weeks before birth. In the case of birds, bird ontogeny repeats dinosaur phylogeny.

Conan Doyle's Lost World: "Whether the bird kicked from the back or the front, this claw met its adversary first, and acted with the slashing, tearing qualities of a sharp knife. The feathers, which were quite long, looked more like elongated fronds of grey fern."

Though I was not destined to see rheas on those blisteringly hot, midday meanderings, I did see a number of other species: short-billed canasteros, for example, small birds with erect, rusty coloured tails, that perched on the thorn bushes and observed me gravely as I observed them; and the Patagonian mockingbird, a grey, long-tailed, cocky bird that strutted about on the ground like a pigeon. One magical afternoon a white-throated hawk sailed directly over my head, a beautiful white and black raptor that wasn't supposed to be this far from the Andes.

I took great delight in identifying these and the other birds that flitted by me as I wandered alone among the thorn bushes on these higher steppes. Bird identification is not an exact science, being (for me) part guesswork, part longing, part knowledge and part imagination. In these it was not unlike paleontology. I was accumulating a satisfying list of bird species, and was fairly certain that my identifications were correct, but there was always an enticing grey area between two similar species within which I most often dwelt; the same bird, for example, depending on the light or the angle or the degree of skill or patience in the observer, could be a Patagonian mockingbird the first time I saw it and a chalk-browed mockingbird the next. The difference between them depended almost entirely on the width of the buff-coloured streak over the eye. It was like the difference, in humans, between a woman with brown hair and a widow's peak and a woman with brown hair and no widow's peak. A crucial point to a lover, perhaps, but unnoticed by a casual admirer. I liked the way such distinctions raised the insignificant to the level of a definition, the forced ambiguity of it, even the way the ambiguity was increased by the wretched reproductions in the book, which were as badly reproduced as those Audubon Series cards I used to collect from Red Rose tea boxes in Windsor. But though I enjoyed the sweet ambiguity of hesitation, nailing a species down was a fine, almost defiant, thrill.

THAT EVENING AT DINNER, Merethe asked me about a bird she had seen perched outside their tent that morning. She said it had a dark head, a beautiful blue body and bright green wings. Together we flipped through the book and eventually decided it was a grey-hooded sierra finch, although the illustration for that bird was particularly smudgy.

Merethe was studying for a degree in museology and working part-time at the museum attached to the University of Oslo, the same museum where Jørn was hoping to get the only full-time paleontologist's job in Norway. For her thesis, she had conducted a study of museum-goers' responses to those little cards of information placed beside exhibits; she wanted to know if anyone read them. Apparently, hardly anyone does. She interviewed individual museum-goers as they were leaving the museum, asking them questions based on information contained on the cards: Did you see the *Tyrannosaurus rex*? Yes. What period did it live in? I don't know. That sort of thing. "Adults with children remembered a bit more," she said, "because they usually read the cards to their kids. But when they're alone, or when the children are alone, all they want to do is look at the exhibits. They don't care much about the science behind them."

I wondered if she had shared her findings with Don Lessem, but she hadn't. It was probably just as well, because the other depressing thing about her findings, she said, was that they seemed to confirm the direction being taken by most major museums in the world: to attract more people through the turnstiles, they were mounting more entertaining exhibits. "Edutainment," they call it, or maybe "entercation." They were spending money on exhibits rather than on original scientific research: who needs science when people obviously prefer pseudoscience? When I was working for magazines, I'd noticed that whenever publishers wanted to make money and still be able to lay claim to a distant relationship with journalism, they started talking about "advertorials." They'd let an advertiser, say a pulp-and-paper company, buy ad space in the magazine and use it to print something that looked like a real article, pointing out for example how environmentally responsible the pulp-and-paper industry was. It could mention how many jobs the company created last year, and how complying with too-stringent environmental regulations threatened those jobs. I always hated advertorials because they were the opposite of journalism, but disguised as journalism. Just as edutainment in museums was the opposite of science disguised as science. Showing models of Steven Spielberg's idea of what a raptor looked like is not showing science. It might get more gawkers through the turnstiles, but is it the responsibility of museums, or journalists, or educators to merely give people what they want? Isn't that what the entertainment industry is for? Doesn't it have to be someone's job to give people what they need, whether they want it or not?

"I like the old museums," sighed Jørn, "the kind where they just stuffed everything into rooms and let people wander around and gape."

Like Rodolfo's museum in Plaza Huincul, I thought. Dust was an essential component of a good museum. Nothing collects like dust. A museum should contain objects frozen at the moment of decay, snatched from the jaws of maniac time at the very second they were being nibbled to dust, particles of it still clinging to their outer surfaces, left there to remind us that this was what happened, this was real life. In fact, there should be a Museum of Dust somewhere, containing vials of dust from all past ages. Dust from the

Egyptian pyramids, from the catacombs of Rome, from the top of a book in a library that had not been checked out this century, from the moon, from distant stars. Dust from Greenland ice cores, dust from the cranium of Peking Man. Dust does not always imply neglect; it can be deliberate dust, cosmic dust. A museum should be like the inside of a very fertile mind, which is inseparable from memory, where things simply gather and are left lying about, ready for another fertile mind to come along and make sense of it. Like Phil's lab in the Tyrrell. Or, I told Jørn, like Dong Zhiming's office in the Institute of Paleontology and Paleoanthropology in Beijing, where Zhiming has amassed not only fossil bones from every region of China, gradually disintegrating to dust, but also ancient, ribbonless typewriters, a broken clock, pencil stubs, yardsticks, a length of watch chain he found in the Ordos Basin and believes once belonged to Teilhard de Chardin, a letter from his mentor, Yang Zhong-jian. Such museums don't even need to exist, I said; they could be thought museums, like Thomas Browne's *Museum Clausem*, or Borge's library of Babel, the items in them preserved entirely in the living memories of their conceivers.

"You could read the cards if you wanted to," said Jørn, "but they would have to be typed on an old typewriter, and stuck to the display cases with thumb tacks or old Scotch tape. And the words on them would be just as nutty as the exhibits. My favourite museum in the world is the Beijing Museum of Natural History. The second floor is devoted to the human body: there's the usual series of jars with deformed human foetuses in different stages of development, or with various birth defects, but there's also a tall, glass case filled with formaldehyde that contains a human, a real, dead man, with his head screwed to the top of the case and his skin peeled off to display the muscles. School kids line up to see this. In another case a woman is lying on her back, her chest cavity opened and all her organs pulled out and labelled. Kidneys, liver, pancreas. It's fabulous."

There were no televisions in the Chinese museums, he said. In Europe, museums are developing life-sized dinosaur robots, thirty-five-metre sauropods, that will be controlled from some observation deck somewhere. The robots will walk freely about the museum, and at certain times they will be made to stand up on their hind legs and eat the tops of potted palm trees. It's hard to know what educational value this will have, but they will be entertaining. As we watched water drip from the ceiling of our kitchen tent onto a box of biscuits, though, we thought that with the money they will spend to construct one dinosaur robot, they could fund expeditions like this one for the next fifty years.

"HOW MANY SPECIES of birds have you identified so far?" Eva asked me in the quarry the next day.

I told her I had about seven, but that there were still quite a few I hadn't been able to identify. Maybe a dozen, altogether.

"Any raptors?" Phil asked. I told him about the hawk and the *carancho*. He nodded: "I'd be interested in knowing the ratio of raptors to other species," he said, "if you're keeping a list."

I should have known Phil would be interested in the raptors, the theropods of the modern bird world. Later that day, when we were working in the far corner of the quarry, Jørn told me that it was Phil who had suggested his Ph.D. topic, which was the lower mandibles of theropod dinosaurs. "In fact," he said, "it's because of Phil that I'm in paleontology at all." When Jørn was a geology undergraduate in Oslo, his grandfather had given him $2,000 to get his driver's licence. Instead, he had used the money to fly to Alberta, where he joined a Field Experience team in Dinosaur Provincial Park. "Phil came down to the park for only one day," Jørn said, "and I talked to him only very briefly. But I went back to Oslo in the fall and switched my major to paleontology."

I asked him what Phil had said that had affected him so deeply.

"It wasn't so much what he said," Jørn told me. "It's just the way he's interested in everything, how he connects everything together. Most scientists are extremely focused on one narrow aspect of their discipline, much as a forester might know everything there is to know about one particular type of tree and nothing at all about other trees, or for that matter the animals or the plants that also make up the forest. Phil sees the whole forest, all the time. For him, I sensed that paleontology was not just the study of some old bones; it was the study of life itself, of species interaction, of ecology and evolution. How dinosaurs lived on this planet a hundred million years ago is inseparable in his mind from how we live on this planet today. That's why this bird thing is so important to him. This link between dinosaurs and birds is really the link between the past and the present."

THE RAIN FINALLY trailed off in the afternoon, the wind dried the mud, and eventually it was possible to make it down the hill without sliding. It was nearing the middle of April and getting colder at night, although the days were still warm when the sun shone. Paul and J.-P. went out geologizing, I did some more birding, and we all kept digging in the quarry. So far we had moved a lot of rock but found very little bone. Phil had part of a skull, and maybe a metatarsal. Rodolfo had his leg bone from last year. The rest of us kept turning up rock. I found a piece of something that was so weathered and disintegrated it looked like a row of rice grains, not worth keeping. The lack of good bone was beginning to get to us. No one mentioned it, but it was on everyone's mind. At one point Rodolfo looked up and said: "This is so *boring*."

Phil had abandoned his awl and was whacking away at a chisel with a large hammer, driving the point four inches into the ground and working it loose by whacking at the side of it, pulling up large chunks of rock but apparently not worrying that he might be driving the chisel into bone.

Eva looked at him in surprise: "Giving up, Philip?" she asked.

I was digging a hole directly across from him, getting deeper and deeper in gravel. I seemed to have found the slump hole, a gravel-filled well surrounded by sandstone. The gravel in it was almost loose, which made for easy digging, but I had the distinct feeling that I was wasting my time. Paleontologists say that quite often they can tell when they're getting close to bone, that they have a kind of sixth sense. It was a bit like divining water. No one was getting that premonition. We were tired and cold and hungry and discouraged. Then the sun came out and warmed us, and at least we were not wet. But we kept the tarp over the bonebed, just in case.

Just before dark, Paul came into the quarry and asked Phil if he would come and take a look at something.

"Let's go," Phil said, almost gratefully, and taking that as a general invitation I put down my tools and went with them. Rodolfo also joined us. We climbed the short cliff to the ridge above the quarry and headed off towards a series of higher cliffs about half a mile to the south. We were crossing a sort of plateau or steppe, the flat, gravelly shelf thick with thorn bushes and goat droppings where I had often gone to look for birds. The low, evening sunlight slanting from behind our backs stretched our shadows out twenty feet ahead of us, as though they were tugging impatiently at our feet.

As we walked, Paul explained that he had seen what he thought might be dinosaur footprints. We quickened our pace. A trackway would be a marvellous find. As Sherlock Holmes knew, footprints tell stories. I thought of the hominid trackway in Africa, which clearly preserves the footprints of two adults and a child crossing a plain nearly five million years ago. Their footprints were pressed into several inches of volcanic ash. In Claudia Casper's novel *The Reconstruction*, the narrator, a sculptor commissioned to make a three-dimensional exhibit for a local natural history museum, tries to catch that magically articulate moment when the footprints indicate that the female deviated from the path, turned and looked back, perhaps to call to the child, perhaps to look wonderingly at the blood-red fire oozing from the mouth of a distant volcano. There's also a dinosaur trackway in Texas that shows two large theropods trailing a family of sauropods across a mudflat: you can actually see when one of the carnivores makes a strike at a dawdling member of the herd, perhaps a sulking juvenile or a faltering grandparent. Once, when walking in the bush behind a friend's house, we came upon a snow-covered clearing, pristine in the evening twilight except for a faint series of brush strokes crossing the centre and ending at a disturbance in the snow flecked with a few spatterings of blood: an owl had swooped low and grabbed a rabbit while it munched on a wild raspberry cane. We saw no owl and no rabbit, and yet their story was clearly written in the snow as in an Aesop fable.

Phil began his career as a paleo-birder with a trackway. From 1976 until 1979, he spent the best part of each summer in the Peace River Canyon in northeastern British Columbia, trying to salvage as much information from slabs of fossilized footprints as he could before the entire canyon was flooded by a hydroelectric project. It was just after the OPEC oil scare, when everyone thought getting off oil and out of the clutches of the Arab cartel was a good idea at any cost. The cost in this case was an unprecedented dinosaur trackway, the first discovered anywhere in North America, containing more than 1,700 footprints formed during the Early Cretaceous. Most of the tracks had been made by two dinosaur species: a carnivore and a hadrosaur, the hunter and the hunted. But in the last year of the project Phil discovered a ten-metre slab of smooth siltstone that had fallen from the side of a cliff and split in two when it hit the bottom. On it were two hundred footprints that "differed profoundly" from the others he had seen. "They closely resemble the footprints of modern paludicolous birds," he wrote in

1981, after he had exhaustively studied the material. Paludicolous means swamp-dwelling; the slab of rock had once been under several inches of water, the mud was closely rippled, filled with root traces and worm burrows, and the prints resembled those made by modern sandpipers. The two hundred tracks, Phil determined by careful measurement, were made by four individuals, apparently ambling around in the shallows pecking at worms, like sandpipers at the edge of a pond, or pigeons in a plaza pecking at corn. Such a glimpse into a few moments of time[2] is still the only evidence we have not only that birds lived in North America during the Early Cretaceous, but also that they lived pretty much as birds live today, something that a pile of bones in a quarry rarely reveals.

After walking for fifteen minutes, Paul stopped and pointed to the ground. We were standing on a flat, bare stretch of sandstone about the size of a tennis court. Scattered here and there in the stone were a number of large, round impressions, each two feet across and half an inch deep, not clearly defined as footprints but rather muffled, as though made with the end of a log wrapped in a quilt. Phil and Rodolfo dropped to their hands and knees and peered into one. Then Phil stood up and paced the length of the exposure. He tried stepping from hole to hole.

"They do seem to follow a track pattern," he said.

"I thought they looked more like underprints," said Paul, meaning that the prints had actually been made in a layer of mud above this layer of sand, but by a creature so heavy that its prints had registered in the sand as well as in the mud, only not as sharply. Later the mudstone layer eroded away, and these under-impressions in the sandstone were all that remained. Still, neither Phil nor Rodolfo seemed very excited. Even if they were dinosaur prints,

---

2 Outside my window as I write this, two mourning doves have landed in my yard and are pecking at seeds in the grass. I timed how long it took one of them to take fifty steps: forty-five seconds. Thus the Peace River trackway perhaps records about one minute of time that passed 120 million years ago.

they were too vague to say anything about the animal that made them. They could have been made by anything heavy. But it was a fine evening, the rain had stopped, and we were away from the quarry. The view from the steppe was magnificent, back over the river valley and across to the hills on the other side.

"Has anyone done any prospecting over there?" I asked, pointing across the valley. I suppose I was thinking that if this quarry didn't start producing bone soon, some of us might begin prospecting across the river.

Phil seemed to be reading my thoughts. He said that last year they had walked the length of the valley at the base of the hills and found nothing of interest. Paul said there was a hill nearby with some early mammal teeth and some fossil wood, but nothing from the dinosaur era. Rodolfo lit a cigarette and kicked at a round stone at his feet. It rolled across the sandstone like a billiard ball and dropped into one of the large impressions.

"I get a lot of people bringing these things into the museum," he said. "They think they're dinosaur eggs. They're just concretions."

As the sun dropped below the hilltop to the west, a red stain spread across the sky towards us, like watercolour soaking into fibrous paper. By the time we were back at the quarry it was dark, and Daniel had started the *asado*.

CHAPTER EIGHT

The Living Screen

T WAS NEARING THE end of April. Late fall was turning perceptibly into early winter: the afternoons were still warm, but there was ice on the ground every morning and the sunsets were calling a halt to work in the quarry earlier and more spectacularly every evening. The Southern Cross continued to announce the nights; when we saw it in the sky we knew the temperature was about to drop twenty degrees and Orion would soon be standing on his head in angry protest.

We were approaching the final, most nerve-wracking stage of the dig. The discouraging phase was over, and bones had been popping up steadily for the past few days—the count was now over thirty. We had evidently reached the ancient surface of the riverbank on which the bones had originally been deposited. They were strewn about haphazardly, totally disarticulated, as though the carcasses had been torn apart by ravaging beasts, although in this case the scavenger was probably Time. Our excitement mounted as each bone was exposed. Sometimes they would be lying on top of one another, a jumble of death. I was working on a rib that seemed to be disintegrating into sand, but beside me, in a deeper part of the quarry, Mike and J.-P. had found the beginnings of something larger and interesting, a femur or tibia lying aslant over a welter of smaller bones. It was as though they were excavating a pile of pick-up sticks: every time they cleared a section of one bone, another showed up that had to be dealt with first. Separating them would be a tricky business, and, with less than a week left, there was no time for tricky businesses. Eva, too, was working on a heap of mixed bones, most of them smaller and from the front of the animal; they had names like lacrimal, jugal, pterygoid.

"Most parts of the skeleton are now represented," she wrote in her field book, "and all can still be attributed to a new species of carcharodontosaurid." This definitely linked these dinosaurs to *Giganotosaurus*, and South American theropods to their African cousins, the carcharodontosauridae, or "shark-toothed reptiles."

So far, everything was still about eight per cent smaller than *Giganotosaurus carolinii*, but Phil seemed confident that we would find something bigger. He said that the small, isolated bone that Thom Holmes had been working on, which he had originally thought to be a jugal, had turned out to be a quadrate. He had also discovered that the bone was pneumatic—hollow and once containing air sacs that had made it lighter (an important feature when the skull was as big as that of *Giganotosaurus*). This was also a feature that linked the large theropods with birds.

Birds have a modified reptilian respiratory system. Reptiles, probably including non-theropod dinosaurs, have septate lungs: simple sacs with a few walls protruding into them to increase surface area. But they are inefficient for air-to-blood transfer, which is why reptiles are usually relatively sedate, not given to prolonged bouts of strenuous or frenetic activity. Birds, however, are frequently required to fly great distances at high altitudes and against strong winds, and their respiratory system has adapted accordingly. Their lungs are septate like a reptile's, but are connected to a series of air sacs which, in turn, branch out into a network of inflatable arteries, called diverticulae, which penetrate into the bird's hollow bones. The system thus fills their entire bodies with air; flying for them is half floating, half swimming.

Features like hollow or pneumatic bones in theropods are thought to be pre-adaptations for flight, characteristics that evolved in response to some other need—in this case to lessen the weight of the skull—but were there when the urge or necessity to fly came along. Animals with solid bones, even if they developed feathers, would be too heavy to fly. Animals with pneumatic bones could, as Loren Eiseley writes, penetrate "the living screen," could make the Darwinian transition from one habitat to another: "The wet fish gasping in the harsh air on the shore, the warm-blooded mammal roving unchecked through the torpor of the reptilian night, the lizard-bird launching into a moment of ill-aimed flight, shatter all purely competitive assumptions. These singular events reveal escapes through the living screen, penetrated . . . by the 'over-specialized' and the seemingly 'inefficient,' the creatures driven to the wall."

A species never knows what it's going to need to penetrate the living screen. A feature can be neutral or even negative for centuries, and then suddenly become crucial for survival when a species is driven to the wall. Birds were pre-adapted for flight by having inherited feathers and pneumatic bones from dinosaurs. Feathers probably evolved first for warmth, then for display, and only then did they turn out to be useful for flight. Hollow bones were necessary as dinosaurs grew to enormous sizes, or needed to be faster to catch smaller prey. Only later did they become handy places to stuff extra divertimenta. Pre-adaptation was their jumpstart into new life.

ONE DAY WE WERE visited by Roberto Saldiva, a thin, spry man in his seventies who lived near Serra Bajo, the high, red plateau that we could barely see in the distance. A few years before, Don

Roberto had found a skeleton on one of the buttress-like approaches to the Serra; he had dug it up himself and brought it to the museum. Rodolfo recognized it as belonging to a small ornithopod, perhaps a hadrosaur, and had gone back with Don Roberto to the spot where it had lain. There he found four complete specimens, all apparently juveniles, none of them known to be there. Rodolfo was puzzled. Hadrosaurs were North American dinosaurs, and were not known to have migrated into South America. And yet here they were. Serra Bajo quite possibly contained concrete contradictions to some of Bonaparte's theories of dinosaur dispersal between the two continents. How had hadrosaurs managed to get into South America when so many larger species had not?

The next day, while Mike, Eva, Daniel and Christian stayed behind to continue work in the quarry, the rest of us drove to Don Roberto's property to buy a goat and to do some prospecting. We went in two trucks, back along the riverbed to the paved road, right for half an hour, and then left on a small, dirt track marked at the junction by a white wringer washing machine. Don Roberto's *puesto* was a simple rectangular box made of cement blocks, with a tin roof, and surrounded by a wire fence covered with vines that had grown wild and twisted, which made a cool-looking courtyard beside the *puesto's* open door. The courtyard was furnished with several tables and rough benches. Rodolfo got out of the truck, stood beside the gate and clapped his hands loudly but without conviction. There was no response from inside.

"He must be out with his goats," Rodolfo said. There was nothing to do but wait. Don Roberto had come to Plaza Huincul fifty years ago, after completing his military service in the south. He worked for YPF, the government oil company, until his retirement, and then took a lease on this place, which consisted of hundreds of acres of sand and thornbush and one huge, upthrusted mesa. The vegetation was so sparse that a hundred square miles would support no more than eighty or a hundred goats. Rodolfo told us that Don Roberto's wife lived in town, refusing to come out to this cement shack, and that Don Roberto lived here alone and only went into Plaza Huincul occasionally to buy supplies and see his family. One of his sons worked at the museum as a technician, making moulds of *Giganotosaurus* teeth for the department of education. The moulds were made of plaster of Paris and spray-painted black with brown roots, to look like real dinosaur teeth: the department had ordered twenty thousand of them, one for every school in Argentina. We looked over the fence into Don Roberto's courtyard; every flat surface, the white metal tables, two window ledges, the fence top itself, was littered with rocks of various shapes and sizes and provenances; Don Roberto must have learned some geology while working at YPF. There were many fossil bones among the rocks, some vertebrae, a few small legbones, but most were fragments of larger bones, weathered out of the mesa's eroding flanks. The small ones might have been turtle, I thought, or crocodile. They also might have been small theropod or large bird. My brain was becoming tired of analysis. Rodolfo and Phil looked at the bones curiously, but without touching them. I wandered over to a small corral, some distance from the puesto, that looked as though it had been built to hold goats for breeding or butchering: a white goatskin saddle was slung over the fence near a rickety gate, and on the far side was a makeshift coop that at one time seemed to have housed chickens or geese. White feathers were strewn on the ground, trampled under a texture of tiny hoof prints.

After a half an hour, Don Roberto emerged from the thorn-bushes on foot and walked briskly into the clearing, smiling broadly. He had lost a goat, he told Rodolfo in Spanish, and had been out since five a.m. looking for it. He'd found it under a gyre of circling turkey vultures. We climbed into the trucks. Don Roberto sat beside Rodolfo, his palms pressed together between his knees as though in prayer, and directed us along a series of deeply rutted roads, all the while speaking in rapid Spanish. The roads had been scraped out years ago by seismic crews during some government survey or other, and were thus arranged in a grid that might have meant something to someone once but now made getting from one place to another a confusing sequence of right and left turns without any apparent forward progress. Serra Bajo loomed, now on the passenger's side, now on the driver's, soaring hundreds of feet above the flat steppeland without ever seeming to get closer. This was an illusion, we knew, but a powerful one. If we had been travelling through jungle instead of dessicated thornbushes it would have been easier for me to view the high, flat-topped mesa as Maple White Land, the lost world and last refuge of dinosaurs, but I managed it all the same. The mesa rose hundreds of feet above the thorn-covered desert, looking remote and formidable and infinitely appealing. I watched for rheas, naturally, but saw only mockingbirds and small desert sparrows. And a dark ring of turkey vultures on the horizon, still hovering over Don Roberto's lost goat.

Eventually we stopped at the base of a low, rounded hill of black sand which turned out to be the base of the serra. Beside the hill was a deep gully, completely dry, that ran up the side of the serra like a wide escalator, and by which we could climb to the top. About halfway up, the gully split in two, one fork continuing straight to the top, the other angling off to the left into a series of level washouts and slumps. Paul and J.-P. took the angled road, Phil, Rodolfo, Don Roberto and I kept to the right. We soon found that the gully did not, in fact, go all the way to the top, but petered out shortly after the fork, so that we had barely gone another dozen yards before we found ourselves clinging to the exposed cliff face, two hundred yards above the flat steppeland, with another fifty yards or more between us and the top. Instead of climbing straight up, we edged our way along the cliff sideways, curving around and eventually coming to an almost level place that opened like an amphitheatre into the side of the mesa. There was some scrawny vegetation, whipped mercilessly by the constant, sand-laden wind that screamed around the concave chimney and was no doubt responsible for its creation. Don Roberto scampered to the centre of this open area and pointed to the ground. When we joined him, we saw a small, neat heap of dinosaur bones and fossil wood, apparently gathered by Don Roberto on some previous hike and stockpiled there for Rodolfo's inspection. Phil and Rodolfo hid their disappointment well: bones removed from their natural settings were almost useless except as curiosities. Don Roberto had no idea where any individual bone had come from, and if any one of them had been important the skeleton and even the horizon of rock it came from would have remained unknown to science. Phil crouched down and picked a small object from the pile, a fragment of bone about two inches long, smooth and oblong, like a piece of carpenter's pencil.

"Gastralia?" said Rodolfo. Gastralia are separate bones that in theropods protected the belly like an armoured waistcoat.

"Maybe," said Phil. "Have you ever seen gastralia on South American sauropods before?"

"It has been described," Rodolfo said dubiously.

"It could be a bit of ossified tendon from a hadrosaur."

Rodolfo frowned. More troublesome hadrosaur material. Don Roberto was already climbing up the side of the chimney towards a smaller cut that seemed to angle upwards towards the mesa's top. When we caught up to him he was seated on a tiny ledge, his back to the mesa's sandstone wall, his feet dangling over the side, gazing proprietorially out over the steppes. The vultures were below us now; for the first time we could see the Andes, barely more than white-tipped mirages that seemed to hover above the distant horizon, detached from the earth below like petrified clouds, a line of lighter-than-air balloons in the shape of mountains. Through my binoculars I saw whisps of vapour swirling between ragged peaks. When the animals whose bones we had just examined were alive, those peaks had not been there.

Don Roberto was shouting to Rodolfo over the wind: when I looked, he was pointing to a large bone that protruded from the cliff face on the ledge on which we were sitting. Rodolfo took his hunting knife out of its sheath and began scraping away the loose sand around the bone. Phil joined him, and soon both of them were using paint brushes, literally brushing loose sand from what was turning out to be a huge sauropod limb bone. It was too much like the naive scene in *Jurassic Park*, and all four of us began to laugh. Don Roberto had obviously found and exposed the bone before, then covered it up and led us here to "find" it for ourselves. I suspected that Rodolfo had told him to leave the quarrying work to the experts, that bones moved from their original places, or destroyed by amateur preparation, were not useful for science. Within fifteen minutes, Phil and Rodolfo had cleaned the huge limb bone; it was lying completely exposed on the ledge, a dark reddish bone about three feet long, slowly turning grey in the heat of the sun.

"Titanosaur humerus," declared Rodolfo. Titanosaurs were huge sauropods that formed part of the faunal exchange between North and South America during the Late Cretaceous, with sauropods moving north into North America and, as Rodolfo was discovering, hadrosaurs and perhaps smaller dinosaurs migrating south. Sauropods seemed to have originated in Africa and spread from there through India to North America. When they died out in North America, the continent was restocked with titanosaurs from South America: no one knows how, since there wasn't supposed to be a Panamanian land bridge for another several million years. And yet *Alamosaurus* from Texas is closely related to *Neuquenisaurus* from Patagonia. Rodolfo was no doubt pondering this conundrum as he poured preservative over the humerus from a plastic bottle he had taken from his day pack, which also contained his lunch. Before finding the preservative he had pulled out an apple, an orange, a box of Saltine crackers and a tin of Pollo del Mar tuna, looking at each as though vaguely wondering how they had got there and what they were for. Then Don Roberto gave another shout, this time from below the ledge. Rodolfo peered over. Don Roberto was on a second ledge, several metres below ours, holding a fossil vertebra about the size of a medicine ball over his head as though trying to hand it up to us. It was too far from us and obviously too heavy for him, so I climbed down and took the vertebra away from him. Then I struggled back up with it and gave it to Phil and Rodolfo.

"More titanosaur," said Phil. "There could be a whole skeleton around here somewhere."

We peered back over the ledge. Don Roberto was even farther down this time, scurrying around in a kind of scooped-out gully, sticking his hands into loose sand, pulling out pieces of bone and making a pile of them, as though gathering Easter eggs. I looked at Phil. He seemed unperturbed. This was, after all, Don Roberto's property.

"Let's have lunch," he said.

"Don Roberto," Rodolfo called down. "*Almuerzo!*"

WHEN WE MET UP at the trucks again it was four o'clock. Back at his *puesto*, Don Roberto invited us in for maté. We sat in the dusty courtyard while he made the preparations inside. Finally he appeared at the open door and called us in.

We entered solemnly, taking off our hats, and sat around a large wooden table, Phil and Rodolfo at one end, Don Roberto in the middle, close to the stove, and Paul, J.-P. and I across from him. The room was sparse and fanatically neat, a room intended to impress an absent wife. The table was covered with cartoon characters, a Tom-and-Jerry tablecloth on which Jerry repeatedly took gigantic sandwiches out of endless wicker baskets, while grinning Tom anticipatorily licked his lips. On the partition, above a propane cookstove, were two large posters, one depicting a smiling girl embracing a goat, the other a luscious vegetable garden, with rows of glistening red tomatoes and green peppers. Behind Don Roberto was a cast-iron woodstove, and from the wall above it, Betty Boop side-eyed us seductively from a red, quilted tea cosy. When the kettle was sufficiently steaming, Don Roberto removed it from the propane stove and carefully poured hot water into a pewter maté cup, stirred the maté with the *bombilla*, then passed the cup across the table to Rodolfo. He must have performed his sipping and spitting ceremony in private, before inviting us in. Rodolfo sipped with a look of intense concentration meant to convey deep, visceral satisfaction, a devout communicant receiving Holy Eucharist, and handed the cup back to Don Roberto. Then it was Phil's turn. While he drank, Rodolfo raised the question of the goat. Paying for it appeared to be a matter of some delicacy. I gathered that we didn't have much money left, but we could not offend Don Roberto by offering a low price that he, as the host, would be obliged to accept. Rodolfo mentioned the number of people the goat would have to feed, and Don Roberto estimated how long such a large goat would have to be left on the *asador* in order to be properly, but not overly, cooked. A minimum of three hours! So long? *Mínimo!* We will have to gather a great deal of firewood. *Si, mucho!* Would fifty pesos be enough, Don Roberto? *Cinquenta pesos!* Nón. *Treinta!* Thirty? Shall we settle on forty, then? If you insist.

Don Roberto rose to put the kettle on the stove for a second round of maté. With his back to us, he said something in rapid Spanish and laughed. "He says he'll bring it tomorrow," Rodolfo said, "and stay to help us eat it."

WE RETURNED TO CAMP and resumed work. Paul and J.-P. were getting ready to leave a few days before the rest of us; they were going to check some geological records at the University of Cordoba, then fly home from there through Buenos Aires. While he was packing, J.-P. fell into a transport of joy at the discovery of a clean pair of jockey shorts at the bottom of his backpack. He came over to my tent to show them to me. He also gave me some of the fossil wood he had collected; there was a midden of it outside his tent, far too much to take with him. We shared some cookies and a glass of Breeders' Choice, promising to meet at the airport in Buenos Aires, or in Toronto, or in Drumheller later in the season. Paul, perhaps having heard the thonk of cork in the Breeders' Choice bottle, dropped by with his cup. Rodolfo had told them that the University of Cordoba was the oldest university in South America, and that Cordoba was a beautiful city, its inhabitants a perfect blend of Spanish and Amerindian, dark, blue-eyed, intelligent and alert. When they were gone (Rodolfo drove them to the bus station in Plaza Huincul) there was nothing beside my tent but two squares of flattened sand, like gravesites, where their tents had been. Paul's square was trenched around; beside J.-P's was a touching, almost child-like assortment of treasures, picked up during his geological forays: fossil wood, gem-like stones, a sun-bleached ram's skull with two horns curving out of its forehead.

While Rodolfo was driving the two geologists to the bus station, the rest of us continued working the quarry. Although we were supposed to be finishing up, we kept finding more bones. Christian found a tooth, a long curve of black ivory, the fourth so far. Mike had found the tip of one earlier, shattered into tiny fragments, and he spent an entire evening gluing it together. I found a tooth root, which meant the tooth to which it had once been attached had come out of the jaw after death. All four teeth had typical theropod serrations along their cutting edges. Theropods did not have molars, they did not chew anything. They tore off and swallowed huge gobbets of flesh and bone. In the process, they shed teeth like sharks, and were constantly growing new ones. Sometimes, Phil said, they even swallowed their own teeth. He could tell the ones that had been swallowed because the theropod's strong stomach acids had partially digested them.

Just before noon we heard the truck below, and soon after that Rodolfo climbed up to the quarry and pointed to a bone at Phil's feet: "Phil, how long did you say that fibula was?"

Phil: "Eighty-six centimetres. Why?"

Rodolfo: "Are you sure?"

Phil, checking his notebook: "Yes, eight-six. Why?"

Rodolfo: "Because, when I was at the museum I checked the *Giganotosaurus* fibula, and it is eighty-four centimetres."

Phil: "Are you sure?"

Rodolfo: "Sure I'm sure. Which means this baby is the biggest theropod in the whole world!"

Phil, grinning: "I knew it had to be."

Eva, who was working on one of the bones in her pile of mixed material, showed Rodolfo a metatarsal that seemed to her to be extraordinarily big.

"Wow," Rodolfo said, "look at that!" He began stomping around the quarry like a gorilla. "Whoompf, whoompf," he said, "this is one huge dinosaur."

THERE WERE ONLY two days left, we had a dozen bones to jacket and take out, and still we found more material. Mike and I were jacketing a huge chunk of sculpted rock that contained half a dozen bones too intertwined to be taken out separately. First we trenched around the bottom of the chunk until it was pedestalled on a narrow neck of rock. Then we pressed wet paper towels into all the crevices and around all the protruberances, in order to make a smooth surface. Phil, meanwhile, was cutting a roll of burlap into narrow strips about six inches wide and three feet long, using a serrated bread knife that reminded me of the serrations on the *Giga-notosaurus's* teeth. Then Mike and I mixed a bowl of plaster of Paris. When Phil finished making burlap strips, we took them and soaked them in the plaster, then wrapped them around the stone-and-bone sculpture with our hands. It took a lot of burlap. Mike mixed bowl after bowl of plaster. We started at the bottom, wrapping the whitened burlap as carefully as we could and patting it flat with our hands, so that no air pockets remained under the material. Here and there we worked lengths of two-by-fours into the plaster to give the resulting jacket more strength. We worked our way up until finally the sculpture, which had once resembled a terra cotta fig-urine, now looked more like a plaster bust, perhaps the head of a huge, featureless giant, perhaps the reconstructed face of *Giganoto-saurus* himself, after plastic surgery. After the plaster and burlap had dried, we would snap off the neck of rock beneath it and try to carry it to the truck.

When we had finished this jacket, there were still half a dozen more. Eva began to doubt that we could do it, and suggested staying on until the job was completed. Someone suggested working through lunch. Phil groaned: "I'd rather cut sleep than food," he said, but we did both, took shorter lunch breaks and worked on after dark, attach-ing lanterns to our foreheads and wearing heavy coats and gloves against the cold. We'd be in the quarry for twelve hours straight. The dark made the work even more nerve-wracking than usual, because we couldn't see if we were cutting through rock or bone. By this time we were at least a few feet below bone level, and so we assumed we were hacking through nothing but stone, but there was always that nagging fear. Every so often we would stop and train our lamps more closely into the trench, and for a time we would feel better. All the while the tarp flapped over our heads in the wind, a timeless sound, I thought; it could be sails above the deck of a darkened ship, or the leathery wings of pterodactyls coming in from a night raid. The sky was clear, but the wind was shifting around to the east, a bad omen: if it rained, we really would be stuck here for days.

ON A THORNBUSH JUST outside my tent, a small, crested bird was hopping from branch to branch, inspecting each parsimonious leaf for aphids or ants: a tufted titmouse, according to my book. Simpson thought the bird that built its nest in thorn trees, from branches torn from other thorn trees, was called "the *chuchumento*, or what have you," but Rodolfo had never heard of such a bird and in fact said the word itself did not exist. It must have been a local name. Daniel said that the huge, unoccupied nests around our camp were made by the little camp follower that Merethe had seen outside her tent, the grey-hooded sierra finch, which Daniel called the *comesebo*. Although the nests had long been abandoned, these birds were constantly about the camp, flying in low, stalling in the air a few inches above the ground and dropping like stealth bombers over breadcrumbs and other morsels from our table or laps. They were the steppes' version of the house sparrow, except that where the sparrow was drab brown and grey, the sierra finch was bright green and blue. I found it difficult to picture such a small, delicate bird ripping off thorn branches and weaving them into the dense, oblong balls that were wedged into the bushes around the camp. Finches normally make cup-shaped nests lined with soft grass. These were like balls of barbed wire, with entrance and exit holes at each end, not unlike magpie nests. I thought they were more likely made by the white-throated cachalote, a reddish-brown bird with a high crest said to make "enormous nests" in the Andean and Patagonian steppes.

On our last full day in camp, I made up my list of birds and gave it to Eva to file in her field report: fifteen species in all. Four predators, two of which—the turkey vulture and the *carancho*—were scavengers.

THE MOOD IN CAMP on our last night was damp and despondent. Daniel made the usual *asado*, we ate it with the usual crackers and *dolce de leche* and drank the last brick of white wine. After the meal we stood around the fire, staring into the coals, shivering in the cold and the wet, anxiously listening to the wind, which was still blowing from the east. Phil moved to the edge of the canyon and stood looking into the dark void. After a while I joined him. Something flew along the canyon in the dark, just below our feet: a late-returning swallow? A bat? Phil was peering across the canyon and up at the quarry.

"Do you hear anything?" he asked.

I listened. Rain, wind, a gurgle of water, Rodolfo's sonorous laugh. "No, nothing."

"That's what I thought," he said, and headed down the path into the canyon. I followed. At the bottom we leapt across a narrow trickle of water and started the climb to the quarry. "We should be hearing the tarp flapping in this wind," Phil called back. "It must have blown off."

It was a difficult climb in the muck made by the rain, especially in the dark. Phil flicked on his headlamp and I had a small pocket torch, but they were mere pinpricks of light in a vast, darkened theatre. Just before we reached the quarry I looked back down at the camp, at the tiny, flickering fire with black shadows moving around it, a medieval scene. The quarry was uncovered and eerily quiet; the tarp, having broken loose from all but one of its anchors, was lying in a corner, looking crumpled and apologetic. Twelve white-jacketed bones lay about on the ground, where they had lain for the past eighty-five million years, softly phosphorescent in the cloud-obscured darkness, like small mummies or giant cocoons. Phil and I pulled the tarp back over them, tucking them in for the night. Tomorrow they would be going on a long journey.

PART THREE

The Badlands

CHAPTER NINE

The Call of
the Wakon Bird

Driving west across northern Ontario in a 1980 Volkswagen camper van is not recommended for the hot-tempered or the impatient. You have to be the kind of person who enjoys endless scenery in the form of rocks and trees, who does not mind being passed by honking, beer-between-the-legs wrestlers in rusty Ford Fairlanes, and who does not panic when, looking in your rear-view mirror, all you see is the word Peterbilt, because an eighteen-wheeler with a Confederate flag on the grille and a load of logs behind it is nudging your rear licence plate, apparently disinclined to gear down and evidently wanting you to swerve immediately out of its way even, or especially, if to do so means to risk plunging a thousand feet straight down into an admittedly picturesque bay of Lake Superior.

It had been six weeks since I returned from Patagonia. After a month decompressing at home in Kingston, tinkering with my ancient but valiant Westfalia van, I'd set out again for Alberta. I'd kept in touch with Phil and Eva: they had stayed on in Argentina to attend a conference in La Plata, then had gone to Italy to look at a 113-million-year-old theropod found there, named *Scipionyx sanmitiucs*, the intestines of which appeared to have been preserved along with its bones. The position of the liver in the abdomen was thought to be not like that of crocodiles, according to its finder, but more like that of birds. Phil looked at it, but decided that the liver in the fossil wasn't necessarily where it had been in life—"Imagine a squirrel run over by a truck," he said—but he did find something other examiners had missed: faint impressions of feathers about the tail. He and Eva were back in Alberta now, getting ready to re-open an *Albertosaurus* quarry they had begun working on the previous summer. Rodolfo was coming up from Plaza Huincul and Jørn was flying over from Norway. July was going to be old-home month, and I wanted to be there for it. The drive was to be my holiday.

After two days, I was still only halfway across Ontario, cresting Lake Superior. "The country on the north and east parts of Lake Superior," wrote the American explorer Jonathan Carver, who paddled around the lake in 1767, "is very mountainous and barren." This was well observed. Carver called the lake "the Caspian of America," because of its size; it was, he said, fed by forty rivers, and so much water ran into it that it was a mystery to him where it all went, since natural evaporation and the shallow rapids at Sault Ste. Marie were obviously far too meagre to account for it. That the lake did not simply keep becoming bigger, Carver speculated, was because the excess water ran "through some subterranean cavities, deep, unfathomable, and never to be explored." He also reported that there was an island in these parts named Mauropas, sacred to the Chippeways, the shores being composed of "large quantities of a shining yellow sand, that from their description must have been gold dust," but that it was protected by a spirit that was sixty feet high, and no Indians would go there or show him where it was. The New World was still relatively new to Europeans in those days, and filled with wonders that could be believed even after scientific investigation.

I was somewhat west of Sault Ste. Marie, closing in on Wawa. Wawa has two distinctions: it is the home of the world's largest concrete Canada goose, and it is known as the black hole of hitch-hiking. All through my youth I had heard about Wawa, about people trying to hitchhike across Canada getting stuck there for days, sometimes weeks, maybe forever. Reportedly half the town was made up of people who had picked up hitchhikers, and the other half were the hitchhikers they had picked up. Wawa was a kind of Pleasantville North, a twilight zone of once-transient souls. Don't smoke up in a cop car and don't get dropped off in Wawa, were the two cardinal rules of hitchhiking in the 1970s.

I therefore passed Wawa feeling a kind of guilty relief. There were half a dozen hitchhikers by the side of the road at the entrance to town, but I didn't pick them up. Who knew how long they'd been there? Some of them had long, grey hair tied in a ponytail in the back; a few were gaunt and unshaven and wore tattered ponchos and droopy-brimmed hats. One woman sat cross-legged on the gravel shoulder, her back to oncoming traffic, her left arm, thumb extended, stretched out and supported on a battered packsack. They all seemed to be gazing wistfully but without expectation into the setting sun. As I approached, hope leapt to their faces. I was bearded and wearing khaki. I was in a yellow, 1980 VW van. I was what they'd been waiting for. Some of them started gathering their things, others stood up and began walking along the shoulder. They knew I would stop. But I couldn't. I tried, my foot actually moved to the brake, I slowed down, but I couldn't stop. I, too, was in the grip of the Wawa legend.

Retribution was swift. Twenty kilometres past Wawa, my engine stopped. One minute I was chugging along as usual, though still spooked by the hitchhikers, and the next I was coasting to a stop halfway up a steep incline cut through the rock of the Canadian Shield. It was as though I had passed through dead air. I wondered if the hitchhikers had put a curse on my engine. I did not even have time to pull off the road. I tried turning the key; nothing. There was almost no shoulder on my side between the pavement and the rock cut. On the other side, a few metres behind me, a narrow lane led off into the bush, with a crude gate across it where it disappeared into the trees. Between the gate and the road there was just enough room for the van. If I let the brake out and allowed the van to roll backwards, across the road, I might make it into the lane before a truckload of logs came hurtling over the hill. I released the brake and the van began rolling backwards, but when I tried to steer I realized that without the key in the On position, the steering wheel locked. With the wheels half turned and the van heading for the ditch, I jammed on the brakes, stopping exactly in the middle of the road, broadsides to anything zooming over the hill. This, I thought, was not good. Freeing the steering wheel and aiming the van roughly at the lane, I got out and began to push, my chin pressed into the mat of dead insects on the front of the van. The van did not, of course, move. I placed my cheek on the insects and pushed harder. The van indicated a reluctant willingness to budge. I could hear the roar of a *Logosaurus peterbiltensis* on the other side of the hill. There was no mistaking its note of joy; it must have caught my scent. Panic sent a charge of adrenalin to my legs, and I pushed harder. The van began to roll, slowly at first, then with some asperity, and finally with positive alacrity, until its back wheels struck a low ridge of gravel left by some prescient grader at the edge of the pavement, and the van stopped, its nose and mine still in the centre of the road. The noise over the hill was definitely coming from a truck that had sniffed the pheromones of terror emitting from my body. I gave a final, desperate heave and the van inched up the ridge of gravel and settled into the narrow ruts of the lane just as the truck crested the hill above us and hurtled past my heels, roaring its fury at its frustrated charge.

I climbed into the driver's seat and stared out through the windshield, considering my position. It didn't seem so bad, I thought. Then I reconsidered. You know you're in trouble when being stuck in a broken-down van twenty kilometres from Wawa seems like an improvement. This occurred to me when I realized that I had no alternative but to hitchhike back to Wawa.

I stood near the rock cut for half an hour with my thumb out, but no one stopped. I was bearded and wearing khaki, and hitch-hiking away from a yellow VW van. Between cars the silence was total, increased rather than broken by the buzz of cicadas and the occasional tweep of an invisible bird. The rock in the cut was ver-tically layered black shale, each layer as thin and shiny as fish scales. I decided to walk towards town, hoping to make it at least to some habitation from which I could phone for a tow truck. I crossed a culvert over a fast-moving, tea-coloured stream named Catfish Creek, and stopped to watch a raven pecking at a patch of fur on the roadside, its bright, yellow eye looking up at me, saying, This might have been you! I soon came to a row of small buildings, with a sign near the road that said "Birch River Cottages. Closed. For Sale," but a car was parked in front of one of the buildings, and I could hear the sound of a lawnmower coming from somewhere behind it. I followed the sound and saw a short, portly man with white hair and a bushy, white moustache pushing a green Lawn Boy over some thin grass behind the cottages. He was wearing ear pro-tectors. I shouted at him, but he didn't hear me. The lawn was mostly sand, and the mower was sending great clouds of dust up to the level of the man's waist. I shouted again and this time he looked up, not startled, turned off the mower, took off the ear protectors, and waited. I went up to him. Van broke down, I explained, point-ing. Need a telephone, call a tow truck. He grunted and started walking towards a building marked Office, which he opened with a key. I followed him inside.

"Place for sale?" I said stupidly. The office was bare except for a telephone, two chairs and a stack of old hunting-and-fishing magazines.

"Worked at the iron mine," he said as he thumbed through a phone book, "and ran this place on the side. Then they shut down the mine." I nodded sympathetically. Wawa was like Drumheller without dinosaurs. "Got a job at the Soo, so now I have to sell. Just as well. No one ever came here anyway." He stopped flipping and ran his thumb across a page. "Put it on the Internet," he said. "Got someone coming tomorrow from Germany to look at it. Lucky I was here."

I agreed with him. He dialled a number and I waited as he listened to an answering machine and then left a message. Then he hung up and dialled another number. There was no answer. He dialled a third number. Then a fourth. He called the OPP. He called three places in Wawa. He called a place in Thessalon and another in White River. No one had a tow truck. He called a radio station, the fire hall, the tourism office, Canadian Tire, the GM dealership and a friend who used to be a mechanic. Nope, nope and nope. Then he called the first number back and left another message.

"Best I can do," he said. "He may check his messages, he may not. He may not check them till Monday." This was Saturday. "Go on back to your van and wait."

"Well, thanks for your trouble," I said. I took out my wallet and handed him five dollars. "For the phone calls," I said.

"Naw, keep it," he said, smiling for the first time. "You're in enough trouble already."

AS IT TURNED OUT, he was wrong, for half an hour later a tow truck pulled up. A young man got out and introduced himself as Derek. He hitched up and towed me into Wawa, deposited me and the van in the parking lot of Canadian Tire, which was closed, and gave me the number of a mechanic named Robert who might be persuaded to take a look at it before Monday. I called the number from the Embassy Restaurant, and Robert said he'd come over first thing in the morning. He was sorry he couldn't work on it himself, he said, because he was going fishing, but he would tell me what was wrong with the engine and I could buy a part when Canadian Tire opened at noon and fix it myself. I said that would be all right. Sunday morning fishing, he gave me to understand, was not something to be interfered with; it was bigger than he was. I thanked him and went back to my table and ordered dinner, vaguely wondering why a restaurant in Wawa, Ontario, would be called the Embassy, but not thinking too deeply on the matter. I felt good. Three people had gone out of their way to help me. I'd lost a day but it could have been worse. If the *Logosaurus* had been a few seconds faster, for example, I would have been a road-pizza. At the table next to mine, eight American fishermen were talking about their day. They were from White County, Indiana. I'd driven through White County, Indiana, just the year before, and I remembered seeing signs along the road that said: "Keep White County White." These men did not seem to be the kind of men who would put up signs like that, but maybe travelling had broadened their perspectives. "Back home," one of them was saying, "if we caught this few fish we'd complain to the lodge owner. But up here, no one seems to care. Up here," he said, looking around the table for confirmation, "it seems more like family."

After dinner I strolled down to the small lake on the shore of which the town of Wawa nestles, and watched loons dive beneath the still surface, trying to guess where they would come up again. Loons are among the most primitive of birds, retaining many characteristics also found in the toothed Cretaceous bird *Hesperornis*. They float almost entirely underwater with only their reptilian heads protruding above the surface. And when they dive, they dive not for weeds, like ducks, but for fish. Carver noted that they were "exceedingly nimble and expert at diving, so that it is almost impossible for one person to shoot them. It takes three persons to kill one of them." But he neglected to say how this feat was accomplished. There was also a huge osprey nest in one of the tall trees visible from where I sat, but I saw no osprey. Carver reported that ospreys, which he called fish hawks, caught fish by means of a secret charm in the form of an oil that they kept "in a small bag in the body, and which nature has by some means or other supplied him with the power of using for this purpose." The osprey would fly low over the water, sprinkling some of this magic oil, which would draw the fish up to the surface, where it could see and catch them. "It is very certain that any bait touched with a drop of the oil collected from this bird is an irresistible lure for all sorts of fish, and insures the angler great success." I thought I'd ask Robert about that in the morning.

By far the most mysterious creature described by Carver in these parts was a magnificent bird he called "the wakon bird," which he said "appears to be of the same species as the bird of paradise." This seemed odd to me, because I'd always thought the bird of paradise was a flower. Was the wakon another mythical bird, like the phoenix, which lived five hundred years in the desert and then, concentrating rays from the sun like an avian magnifying glass, caused itself to burst into flames, and then rose from them to live another five hundred years? Or the Thunderbird, which some believe was the precursor of the modern condor, now extinct but still extant in the living memory of the West Coast people? The Indians, said Carver, venerated the wakon "for its superior excellence," and called it the bird of the Great Spirit, Kitchi Manitou. "It is nearly the size of a swallow, of a brown colour, shaded about the neck with a bright green, the wings are of a darker brown than the body; its tail is composed of four or five feathers, which are three times as long as its body, and which are beautifully shaded with green and purple. It carries this fine length of plumage in the same manner as a peacock does." This might have been a partridge of some kind.[1] Carver never saw the fabulous Wakon bird, but was assured of its existence.[2]

I resolved to defy the Wawa legend. When I had fixed the van the next day, when I was leaving town, I would pick up a hitch-hiker, maybe all the hitchhikers, and take them at least as far as Thunder Bay. I felt an enormous surge of self-importance, of the kind often mistaken for virtue, as I pictured myself stopping by the roadside, sliding open the side door of the van to a group of these tattered pilgrims, my fellow travellers, and of the tales we would tell as we winged our way westward. I would tell them about Patagonia and Che Guevara. One of them would have a guitar and we would sing "Four Strong Winds," and they would marvel at the news that Ian and Sylvia had split up, and that Ian was still in Alberta, singing cowboy songs. They would tell me of the sturgeon fishing, of the closing of the iron mines, and the building of Wawa's famous, con-crete Canada goose.[3]

However, when I had fixed the van the next morning (all it needed was a new coil) and reached the intersection where the road into Wawa met the TransCanada, there was not a hitchhiker in sight. The ponytailed and ponchoed pilgrims had vanished, as though into thin air. Had they already been picked up, I wondered, or did they take Sundays off?

1 Samuel Hearne, in his Hudson Bay journal for the year 1772, noted that the wood partridge was "a handsome bird: the plumage being of a handsome brown, elegantly spotted with white and black. The tail is long, and tipped with orange." This was the spruce grouse, though, which elsewhere he said was less beautiful than the ruffed grouse, and neither of them was called the wakon. The name for the partidge in the south, he said, was Mistick-a-pethow, and the ruffed grouse was the Pus-pus-kee.

2 I have not seen any other reference to the mythical wakon bird, but the Sioux word for Great Spirit is "wakan," also spelled "wakon," and William Henry reports in his *Travels*, published in 1776, that the Assiniboines "believe in spirits, gods, or manitos, who they denominate wakons." Carver was never anywhere near the Assiniboines. He was, however, accused of having cribbed many of his descriptions from earlier French reports, and this may be an instance of it.

3 "Wawa" is the Ojibway word for "wild goose." In the 1950s, there was an attempt to change the town's name to Jamestown, in honour of Sir James Dunn, the mining magnate who owned the local iron mines, but the citizens rebelled and the project was dropped. Fortunately, for iron mines come and go, but Canada geese are always with us, especially concrete ones.

TO MAKE UP FOR lost time, I drove 650 miles that day, stopping only for gas and to make tea and a sandwich at a roadside picnic spot overlooking the shining waters of Lake Superior, and finally pulling into a provincial campground halfway between Thunder Bay and the Manitoba border at ten-thirty at night. I had reached that stage of exhaustion in which it took several minutes to decipher the meaning of two mysterious red dots on the highway that kept moving from side to side, vanishing altogether, then suddenly reappearing, until I realized with a start that they were not UFOs but simply the tail lights of a car ahead of me as it went over hills and around curves. I awoke in the morning to the hopeful sound of birds, and as I drank coffee sitting at my picnic table I counted five species of warblers in the pale-leafed birch trees that separated my campsite from Sandbar Lake: magnolia warblers, a northern parula, two black-throated greys, a Canada warbler and an American redstart, tiny, beautifully marked birds that drifted from east to west through the trees with the advancing morning sun.

The highway began to climb up to the Prairies. At first I thought it was an illusion, an impression of endless climbing caused by the fantastic heights achieved by the mountainous terrain that separated Lake Superior from the Hudson Bay lowlands. The Prairies are an ancient seabed; how could they be higher than these precipitous cliffs and vertiginous drops into the silvered waters of the lake, which rippled and glistened below me like the scales of some enormous reptile in the blinding northern sunlight? But it was true. Sault Ste. Marie is only six hundred and thirty feet above sea level, and Thunder Bay is even lower, at six hundred and seventeen. But Kenora, Ontario, to which I was now headed, is eleven hundred feet, and the bottom step of the post office in Dryden, where I stopped for lunch, is twelve hundred and twenty-four feet above sea level. I was definitely climbing.

I walked around downtown Dryden looking for a bookstore and finally found one on a side street, but it didn't have many books in it. A few bestsellers and the latest mysteries, propped up on tables and facing out from shelves, to make the store seem better stocked than it was. I bought a Martha Grimes and some writing paper, and went back outside. On a bench across from the largest pulp-and-paper mill I had ever seen in my life, I ate an apple and wrote a letter to my wife. My attention was soon distracted by a flock of ring-necked gulls wheeling above the effluent pond that drained out of the mill. The pond looked surprisingly clean, but I doubted that there could be anything living in it that gulls would want to eat. But then gulls want to eat any number of things, not all of them living. The mill belonged to Weyerhaeuser Canada, a branch of the American forest-products company that had just bought Macmillan-Bloedel. A banner stretched across the front of the building read: "Safest Mill in Canada!" and a wooden sign erected beside the mill's guarded parking lot added the information that

THIS MILL HAS HAD

_5_ DAYS

WITHOUT A REPORTABLE ACCIDENT.

DRIVING OUT OF THE mountains of northern Ontario onto the flatlands of eastern Manitoba was like waking from a dream in which I was being chased by monsters to find myself lying on the floor of a strange house. The narrow, twisting, treacherous road expanded into a twinned, arrow-straight, four-lane superhighway, making it easier to negotiate the level, boring expanse of unbroken plain. To keep myself awake I began to take note of the weird brand names we have given to our cars and trucks. There was a time in our history when most of our vehicles were named after animals, so many that it is possible to draw up a list of species and speculate on their distribution and meaning, which I did as I was figuring out how to bypass Winnipeg. Among the domesticated animals are the

Bronco, the Pinto, the Mustang, the Colt, the Taurus and the Ram. These, with the exception of the Pinto, are the muscled vehicles, useful, maybe even recreational, but not something you'd take to the opera. Wild animals are more numerous and diverse, as in nature, but here's a curious thing: the prey species, such as the Rabbit and the Impala, are far outnumbered by the predators, the Lynx, the Bobcat, the Cougar and the Fox, an imbalance that would never be found in nature, at least not for long. Birds are represented mainly by raptors—Thunderbirds, Eagles, Falcons and Tercels—with only one songbird, the Skylark. Fish are predominantly the hunters rather than the hunted: Marlins, Stingrays and Barracudas. The only snake is the Cobra. This is not an ecosystem with a future. There are many non-vicious snakes in North America, in fact most of them are relatively benign, but I don't suppose anyone is going to drive around in something called a Pink-Belly or a Black Chicken. There are, however, a whole class of snake called Racers, and I would have thought there'd be a list of promising names in it: what's wrong with the Indigo Racer, or the Coachwhip? We seem, however, to favour predators, the very animals that, as a society, we have relentlessly hunted down and removed from our midst. We killed off the cougar and immortalized it with the Cougar. And the trend remains if you extend the list to include other vehicles, as I did as I turned off the TransCanada and headed down to Steinbach, and began to pass huge dealerships displaying massive farm machinery, recreational vehicles and construction equipment. We have replaced sled dogs with Arctic Cats; we scrape the earth with Caterpillars; we turn prairie grassland into wheat fields with a tractor that runs like a deer. For a while, anyway, we named our vehicles, like our subdivisions and our shopping malls, not for what they were, but for what they destroyed.

At Steinbach, I turned west and then south to La Rochelle, then west again to the Red River, a straggling, muddy stream that seemed hardly capable of the damage it caused along this stretch with its regular flooding. Then south along Highway 75. I intended to drive straight west, on Highway 14 in Manitoba and 3 in Saskatchewan, until I hit the Alberta border, then angle up to Drumheller. But I had given myself extra time for side trips, rest stops and breakdowns along the way.

EXCEPT FOR A FEW Triassic bones in the Arctic and some Jurassic theropod footprints in Nova Scotia, most of the dinosaurs found in Canada lived in Late Cretaceous Alberta, a period that extends from about 100 million years ago to 65 million years ago, and hadrosaurs are the most common among them, in fact are the most widely dispersed family of Late Cretaceous dinosaurs in the world. Most of the dinosaur skeletons Phil saw as a child in the Royal Ontario Museum were hadrosaurids, the duckbills and tubeheads found in the 1920s by the Sternbergs. For every theropod skeleton that turns up in Alberta's Dinosaur Provincial Park, there are at least twenty hadrosaurids. There are places in the park where you cannot rest your eye without it lighting on a fragment of hadrosaur bone, part of a rib, a shattered thighbone or a vertebra sticking out of a coulee wall, and unless it is an exceptionally complete skeleton, you don't bother collecting it.

Although they're called duck-billed dinosaurs and classed among the ornithopods, or bird-feet, their feet weren't very bird-like and their jaws had cheeks with large, flat, self-sharpening and replaceable teeth in them. I've always thought of hadrosaurs as the moose of the dinosaur world. They had the same sloppy, side-to-side-chewing mouths and, at least in the way they've been depicted by paleoartists like Greg Paul, the same dozy, myopic look. I can picture one standing belly deep in a swamp, some kind of spinach-like fern hanging from its mouth, looking blankly at me as I pass on a trail. Or with its back to a giant cypress tree, fighting off a pack of snarling, wolf-like theropods with its bluntly clawed forefeet; one of the theropods has locked its teeth on the hadrosaur's nose and is kicking at its throat with its hind claws, and the doomed beast is vainly trying to shake it loose. Bleeding from the nose, panic in its eyes, it decides to make a run for it, and that's the end of it.

The first dinosaur bone found in western Canada was a hadrosaur thigh-bone, found in 1875 in what is now southern Saskatchewan by George Mercer Dawson. Dawson was the official naturalist of the British North American Boundary Commission, a sort of United Nations of unlikely expedition mates that had set out the year before to survey and mark the 49th parallel, recently established as the boundary between Canada and the United States, from Lake of the Woods to the Pacific Ocean. Half the crew was British and the other half American. The British had hired a group of twenty Métis guides, and the Americans brought along two companies of cavalry and another of infantry—230 men in all—for protection, they said, from the Indians but also possibly from Dawson's Métis. It had been Métis rebels under Louis Riel that, only three years before, had prevented earlier survey gangs from dividing the Prairies into neat little sections ready to receive farmers from the east, and were preparing for continued resistance farther west, where the commission was going. At any rate, when there was converse with Indians, the army seemed to be elsewhere: at one point the camp was visited by four hundred mounted Piegans, who helped themselves to all the tea, sugar and matches in the supply wagons and left a quantity of leather rope in exchange. There must have been a lot of tension in camp, what with no tea and all that rope. In the inevitable group photographs, no one is smiling. Everyone seems to have worked independently, the astronomers going on ahead to chart the parallel, the surveyors following to mark the line, work crews bringing up the rear to build the cairns, and Dawson left behind to do whatever it was that naturalists did. Mostly he collected angiosperms. His journal entries are full of sketchy botanical information: "White Mud River, July 4, 1874, common grass of drylands everywhere west of about Turtle Mountain, forms short, curly sod." "North crossing White Mud, July 11 and prairie everywhere. The Indian turnip, so called." He was interested in animals, but they were harder to collect. "Arrived Woody Mountain Settlement, everyone preparing to move west into Cypress Hills for winter. Wild oxen so-called found about here. Have escaped from American forts & live on the plains."

But he was trained as a geologist and had a family connection with dinosaurs, for his father was William Dawson, Nova Scotia's provincial geologist before becoming principal of McGill College, in Montreal. William Dawson had acquired the jaw and teeth of what at the time was thought to be Canada's first dinosaur, *Bathygnathus borealis*, a carnivorous creature found in some Triassic beds in Nova Scotia, and had shown it to Sir Charles Lyell, Britain's most famous geologist, in 1852. George Dawson had just graduated from the Royal College of Mines in London, where one of his professors had been T. H. Huxley. During the expedition, he spent a lot of time looking for coal. He had been instructed by the government to assess the West's economic potential, but he also knew from Huxley and Lyell that fossils were often found in coal beds. On June 29, 1875, twenty kilometres south of Wood Mountain, he was digging through a layer of sandy clay that had thin streaks of lignite running through it "with well preserved fossil plants," when he broke through to the underlayer, a purplish grey bed of undulating sandstone. In the lower section he found "bones of turtles? & of some large vertebrate." The bones were encased in extremely hard rock called ironstone. Dale Russell once compared removing ironstone from around fossil bone this way: "Take a soft-boiled egg and drop it into a bucket of cement. When the cement hardens, chip it away without breaking the egg." Dawson seems to have worked hard at it, for he was wrapping up the fossils for shipment by July 11. He sent them to Edward Drinker Cope, at the Academy of Natural Sciences in Philadelphia. Cope identi-fied the turtle bones as turtle bones, and the vertebrate bones as an unspecified species of hadrosaur, which probably enlightened Dawson not at all, and eventually returned them to the Geologi-cal Survey in Ottawa, where they promptly went missing and have not been seen since.

THE CLOSEST TOWN to the spot where Dawson found the hadrosaur is Killdeer, Saskatchewan, on a stretch of Highway 18 that is, for some obscure reason, called Highway 2. This far south, ten kilometres from Montana, the prairie is not exactly flat and not exactly hilly, but gently undulating, like a calm sea before a storm, and here and there is trenched through by riverbeds. Where water collected in the coulees, I stopped to watch small, white flotillas of pelicans preening in the bright sunlight attended by bursts of Franklin's and Bonaparte's gulls. Every now and then I'd catch a distant glimpse of badlands, and imagine George Dawson turning his prairie schooner towards them, taking out his notebooks and botanizing equipment, and scanning the horizon for bison and pronghorn antelopes. I wondered why the town was called Killdeer. There weren't many places named for birds that I could think of. Dunrobin, maybe, and Ravenscrag. Gull Lake. Goose Bay. Phoenix, Arizona. Phoenix reminded me of how hot it was here in Saskatchewan. The sun had baked the clay on the distant hills to a dull shade of yellowish grey, like bread sprinkled with wood ash. There didn't seem to be much traffic, but I was watch-ing for hawks and might have missed it. Before reaching Killdeer, however, I saw a woman leaning against a truck that was stopped in the middle of the highway. When she saw me she began waving her arms and hopping about. I stopped behind her truck and she hurried back to my window.

"Do you know anything about trucks?" she asked.

I had just installed a new coil in the van, but I didn't think that that qualified me as a Class A mechanic, so I said, "A little."

"Mine's stopped. Won't start. See if you can start it for me. I've been here three hours and you're the first to come by."

I said, "Sure," and swaggered over to her truck. The gas gauge showed half full, so I turned the key. The engine groaned but wouldn't fire. "Sounds like you need a new coil," I said.

"Well, are you going into Killdeer? Can you give me a lift? I'll have to phone my brother to come and get me."

"I am," I said, and we got into the van. She was a small woman in her late fifties, and was obviously the worse for her long wait in the sun. Her face was bright red and she was acting a bit oddly. I gave her a bottle of water from my cooler. "You could die of dehydration out there," she said when she had finished it. "I couldn't stay in the truck any more, it was so hot, but then it wasn't much better outside on the pavement. I thought if someone didn't come along soon I'd go into the field and dig a hole, lie down in it and cover the hole with grass." She laughed. "It might not save my life, but at least I'd be properly buried." She told me her name was Marie and that she'd been a schoolteacher for thirty years and was now retired. "They started busing the kids all the way up to Moose Jaw," she said. "Two hours each way. Does that make sense to you?" I said it did not. "I taught Mark Messier," she said proudly, looking at me closely to see if I recognized the name.

"Did you teach him to play hockey?" I asked.

She laughed again. "No, I taught him history."

"Why is Killdeer called Killdeer?" I asked her.

"I haven't the faintest idea," she said.

"Are there a lot of killdeers in this area?"

She shrugged. "I wouldn't know. Maybe it's where the hunters come during deer season. Are you interested in birds?"

"Sort of," I said. I didn't think I'd bother mentioning small theropods.

"Well, you want to go up to Moose Jaw," she said. "After you drop me off. You have to go north anyway, unless you want to go down to Montana."

"Why Moose Jaw?"

"You haven't heard about the burrowing owls?"

"Burrowing owls?" I said.

"Yes. They're owls that live under the ground. Apparently they're getting quite scarce, but they have some in Moose Jaw. At the race track."

"I've never seen one," I said.

"Neither have I," she said. "Mind if I smoke? My nerves are shot."

Killdeer wasn't much more than a few houses, a grain elevator and a Cardlock gas station. I left Marie at a phone booth outside a grocery store and headed north. To Moose Jaw, which I privately rechristened Hadrosaur Jaw.

BURROWING OWLS ARE rapidly disappearing from the face of the Earth. They are small owls that adapted so well to prairie life that they were once almost as common as gulls. Farmers liked them because, unlike other owls, they fed principally on grasshoppers. They hunted during the day rather than at night, and instead of perching and nesting in trees, which were scarce on the Prairies, they took over abandoned gopher holes, of which there were millions. These adaptations have done them in, however, since hunting during the day made them vulnerable to being hunted themselves by human idiots, and eating insects and living below ground made them susceptible to agricultural pesticides; their decline dates from the 1930s and has been in a continuously accelerating downslide ever since. They were listed as threatened in 1979, promoted to endangered in 1995, and a count two years ago produced only six pairs in British Columbia, thirty in Manitoba, eight hundred in Alberta and twelve hundred in Saskatchewan. This year, only fifty per cent of the owls that had migrated south the previous fall returned from Mexico. That left about a thousand pairs all told. The burrowing owl appeared to be another species on its way out. I figured that if I were ever going to see a burrowing owl in the wild I had better get up to Moose Jaw when I had the chance. This was more than just a birder's need to tick off another species, more than the kind of collector's avarice that had little to do with the aesthetic value of the thing collected. To a real bird lover, seeing a bird in the wild is one way of keeping it alive.

The race track turned out to be part of the Moose Jaw Exhibition Grounds, and when I got there the fair was in full progress. I could see the spinning Ferris Wheel and Tip-o-Whirl long before I turned in off the main road and, driving slowly over thick firehoses and thin electrical wires pressed into sodden grass, parked beside the horse barn. One of the side doors to the barn was partly open, and I squeezed into a pitch black, cavernous arena with pinholes in a ceiling so high they looked like stars on a dark night. It was cooler inside, and there was the sweet smell of horse manure and fresh wood chips. When my eyes adjusted to the darkness, I was standing behind two women in cowboy hats leaning on a board railing, watching a young girl on a horse canter around what looked to me like a hockey rink with a dirt floor.

"Are there some burrowing owls around here?" I said, and they both turned around sharply. The brims of their cowboy hats collided.

"Oh my God," one of them said, raising a fringed arm, pointing over my shoulder, and holding it there while she caught her breath. "Back there. Through the parking lot, through a gate, through another parking lot, through another gate, and it's on your left."

The second parking lot was where the carnies had parked their trailers, and I threaded my way through a maze of Westwinds and

Travelbreezes until I was hopelessly lost. After a while a space opened up and there were people sitting around a picnic table playing cards, drinking beer, talking, listening to radios, young men trying to look old, old women trying to look young. Sound pounded against tin and heat vibrated off the pavement. The nearest couple to me was a young man wearing a black leather vest with no shirt and jeans with a belt buckle the size of a hubcap, talking to a thin woman with bright red hair frizzed high above her head and held there with a purple scarf. They stopped talking and looked at me. I had my binoculars around my neck.

"How do I get out of here?" I asked.

"Here," the man said. "Follow me." He stood up and handed his beer to the woman. "Looking for the owls, eh? It's this way."

I followed him through the trailers and eventually we spilled out onto the edge of the parking lot where a wide, Paige-wire gate opened into a grassy field containing a Quonset hut and a dirt track. A sign at the Quonset hut read: "Burrowing Owl Interpretive Centre." I said thank you and mumbled something about probably seeing him again on the way out.

"No problem, man. Stay cool."

When I stepped into the Quonset hut, a young woman of about high-school age came from behind a counter and introduced herself as Natasha. She asked me if I wanted to tour the facility before going out to see the owls, and I thought it prudent to say yes, although I thought I could take in most of it from the door. She led me through an arched doorway into a burrowing-owl burrow, constructed to human size, the walls and ceiling painted to look like the inside of dirt, with leafless vines and roots dangling realistically all around. In one corner, a five-foot adult burrowing owl made of papier mâché stood precariously at the alert, looking a bit like Sesame Street's Big Bird on drugs but with longer legs and painted a drab brown and white. Beside it on the floor were two Styrofoam eggs the size of footballs, representing a nest, and scattered here and there were several apparently real stuffed Richardson's ground squirrels. "They mostly eat grasshoppers," Natasha said apologetically, "but occasionally they will eat mice and small snakes."

"Do they dig their own burrows?" I asked.

"Oh, no," she said. "They take over burrows abandoned by ground squirrels, pocket gophers, badgers, things like that."

"Maybe they should be called borrowing owls," I said.

"Hmm-hmm," she said. "Over here we have a display case showing the various enemies of the burrowing owl." The case, which looked like a coffin with shelves and a glass front, contained several stuffed animals and reminded me of the natural history room in the museum at Plaza Huincul. There was the red fox, but this one also had a coyote, a badger and a coiled rattlesnake. I noticed there were no toy tractors hauling miniature chemical sprayers, or little plastic men in camo jackets on three-wheeled Dirt Trikes with tiny whisky bottles in their hands. There was, however, some kind of speckled hawk nailed to a forked willow branch in the left-hand corner.

"What kind of hawk is that?" I asked.

"A visitor last week said it was a ferruginous hawk."

I'd never seen a ferruginous hawk, and regretted momentarily that I couldn't tick it off my life-list now that I'd seen a stuffed one.

"This," said Natasha, moving me along, "is our man-made burrowing owl nest, which is what these birds are using." In the centre of the floor was a large, wooden box with a length of weeping tile coming out of one end and a round hole in the top. Above the hole was a bottomless plastic bucket with a wooden lid. Apparently the box, with its plastic bucket attached, was buried in the ground, and the owls crawled into it via the weeping tile. Every day someone would come, take the lid off the pail and drop dead mice and grasshoppers into the box for the owls to eat.

"You mean the owls here aren't even using real burrows?" I asked.

"Well, actually, no," Natasha said. "You see, last year we had four nesting pairs in actual ground-squirrel holes, but they didn't come back after migrating. We don't really know what happened to them. So this summer we put in four of these man-made nests and released eight captive-bred birds, which were raised in artificial nests like these, hoping that they would establish natural breeding pairs and produce offspring."

"Has it worked?"

"Not exactly." Natasha looked anxious. "Actually, no, it hasn't worked at all. There are only three birds left. The others were either caught by hawks or coyotes, or maybe they just flew away. Anyway, the three that are left are each occupying different nest boxes and I think they're all males. So," she shrugged. "Maybe more will come back next year."

We left the Quonset hut and walked across the field to a small, enclosed grandstand that was being used as a blind. Natasha had a field telescope already set up and trained on one of the boxes in the centre of the race track. She looked through it, made a few adjustments, and stepped back for me to take a look. "It's there," she said. I peered through the lens and saw a small, rounded head poking up above the level of the grass. It was clearly an owl's head. It was looking straight at me in a curiously accusatory manner, like a cat waiting for its dinner. "There's another one, to the left a bit," Natasha said, and I swung the telescope around and picked up the second owl. This one was standing on a mound of dirt, slightly higher than the first, and I could see its absurdly long legs, like an owl on stilts. Its saurian ancestry was clearly evident in these legs; when I pictured it without feathers I could distinctly make out the lean body of a small theropod, and see it holding down a hapless field mouse with its large talons as it ripped at the flesh with its hooked beak.

"The third one," said Natasha, "should be right over—Oh shit!" she yelled. I looked up in alarm to see her run out of the blind and tear across the track into the field, waving her arms and shouting wildly as she ran: "Hey! Shoo! You leave my birds alone, you!" I looked ahead of her and saw what she was yelling at. A coyote had slipped under the fence on the far side of the field and was sitting unconcernedly in the middle of the race track, at about the halfway marker, watching Natasha flailing towards it across the open field. I looked at it through the telescope. It was a beautiful animal, a light, lustrous brown with black and white markings across its shoulders. "Get out of here, you mangy old coyote!" I heard Natasha yell. "Go on! Shoo!" The coyote yawned and licked its lips. Natasha reached mid-field and the coyote stood up and stretched, its forepaws laid out in front, its hind legs bent at the knees, its straight, bushy tail angled down towards the ground behind it. When Natasha was almost at the edge of the track, it gave a kind of Jesse Owens back-step and, hardly seeming to move its legs, sprang towards the fence and was gone. Natasha stood on the track and shook her fist after it. I admired her courage, and went out onto the field to congratulate her. Together we checked the two remaining nest boxes, but there was no third burrowing owl in either of them. No scattered feathers, either, so I told her I didn't think the coyote had got it.

"Not this time, maybe," Natasha said, looking around. "But the darn coyote might have scared it away for good. The poor thing might not come back. Agh," she said angrily, "if only I'd had a gun!"

I didn't want to make her feel any worse, but my sympathies had been divided during the chase scene. I thought that if they had to build coyote-proof condos for the birds, hand-feed them pre-killed mice every day, and shoo away nasty predators, then burrowing owls were probably not going to stage a comeback. As it was, even though I had now seen a live burrowing owl, it didn't seem any more real to me than the ferruginous hawk had been, or the stuffed rhea in Plaza Huincul. I couldn't, in all conscience, add the burrowing owl to my life list. I still needed to see one in the wild.

But I had seen one beautiful coyote.

A Day at the
Bone Museum

I F SASKATCHEWAN PLACE names are anything to go by, the province seems to have been settled by two kinds of people. There were the dreamers, the optimists, who gave their towns names they then had to live up to: Success, Plenty, Reward, Unity. I thought I would have been uncomfortable in these places. The pressure must be enormous. How could you possibly starve with dignity in a town called Plenty, or fail in the midst of Success? Imagine being impotent in Climax, or saying you came from Renown and hearing someone reply they'd never heard of it. Much safer, I thought, to be simply descriptive. The more down-to-earth settlers just looked around and named their towns after what they saw: Yellow Grass, Goodsoil, Antler, Foxwarren, Snipe Lake.

This thought did make me wonder about Moose Jaw, so I looked it up before leaving town. John Palliser, leader of an exploratory expedition to the Pacific Ocean, camped at the present site of Moose Jaw in September 1857 on the banks of a large creek. When he left, he had a great deal of difficulty getting his wagons down the embankment and across the creek: after several attempts, he gave up and made a wide detour south, to a place where the banks were less steep. The original Cree name for

Moose Jaw means: "The place where the white man mended the cart wheel with the jaw bone of a moose." Talk about descriptive.

Eastend, in the southwest corner of Saskatchewan, is another of the descriptive ones, having taken its name from the unremarkable fact that it is situated at the east end of the Cypress Hills. The town just before it is called Dollard, after Dollard des Ormeaux, the hero of early Quebec; it was given that name by Gabrielle Roy's father, who was in charge of setting up immigrant settlements in Western Canada in the 1920s. And just after it is Robsart, which is a lovely name, although I'm not sure what, if anything, it described. And since it's now a ghost town, there's no one there to ask.

The Cypress Hills[1] are the highest points of land in the province. R. D. Symons, in his book *Hours and the Birds: A Saskatchewan Record*, notes that the prairies are not a single, flat tabletop stretching from the Great Lakes to the foothills of the Rockies, as is usually assumed, but actually proceed westward in a series of giant steppes, with Manitoba, which has an average altitude of about nine hundred feet, being Steppe One, eastern Saskatchewan (fifteen hundred to two thousand feet) as Steppe Two, and western Saskatchewan and eastern Alberta (two to three thousand feet) as Steppe Three. This tilt towards the northeast is why most prairie rivers drain towards Hudson Bay.

The Cypress Hills, straddling the Saskatchewan-Alberta border, are a high bench of land poking up fifteen hundred feet above Steppe Three, and push the water of the Frenchman River south into the Missouri system and eventually into the Gulf of Mexico. On its way, it passes through Eastend. During much of the Late Cretaceous, the entire Prairie provinces were the floor of a vast

---

1 The hills, which are made up of badlands surrounding countless rivers, were named for their evergreen trees by the Palliser Expedition. The trees are mostly gone now, and they weren't cypress trees anyway. The rivers, Bone Creek and Skull Creek, have alluring names for paleontologists but probably refer to bison remains, as the coulees were used as buffalo jumps by the Plains Cree.

stretch of salt water known as the Western Inland Sea, which joined the Arctic Ocean to the Gulf of Mexico. The Cypress Hills were left unscathed by receding glaciers, and they are all that's left of that ancient seabed. They certainly would have been the first points of dry land to appear above the sea's surface when the waters dried up, as they did towards the end of the period. As the edge of the Inland Sea moved from Alberta into Saskatchewan, the dinosaurs moved with it, and in the Cypress Hills their bones are usually mixed with those of marine creatures, such as crocodiles, turtles and Cretaceous shorebirds like *Hesperornis*.

I parked the van in front of a large, red-brick building that looked like an old bank at Eastend's main intersection, locked it, and walked a block south. Eastend had blocks, with stores and sidewalks, a real town with a population of about 650. I stopped at a newly painted, white building with a sign on it that read "Eastend Fossil Research Station," went inside and asked for Tim Tokaryk. Tim, to whom I had written warning of my visit but without specifying a date, was Eastend's resident paleontologist. The woman behind the counter looked up from her novel and said he wasn't in.

"He's in his bookstore," she said.

"Where's that?"

"Right in front of where you parked your van," she replied.

I'd missed the little sandwich-board on the sidewalk by the bank building: "Redcoat Booksellers." Tim and his wife, Norine, had bought the building and were starting a second-hand bookstore in it. They were hoping to open that weekend. I found Tim sitting at a table staring at a computer, surrounded by empty shelving and full cardboard boxes. He looked up briefly as I entered, then went back to the screen and tapped a few times on one key.

"You didn't have to lock the van," he said. "People get insulted."

"Back home," I said, "we say locks only keep the honest people out."

He laughed, we shook hands, and he pointed me to a large, red plush sofa chair. On a small table beside it was a stack of books, most of them about science and paleontology.

"I pulled those out for you," Tim said. "Thought you might be interested."

He was a big, burly man in his mid-thirties, I guessed, although as I get older I find it harder and harder to estimate how much younger than me other people are. His upper front teeth were missing, and I asked him if he played hockey. "Yeah, goalie," he said, grinning spaciously. "How could you tell?" In the late eighties, before coming to Saskatchewan, he'd worked at the Tyrrell Museum, and knew most of the people I knew. "I was a preparator," he said. "It was a great place to work in those days. Everything we found was new. Jane Danis, the museum's collections manager, kept a huge cabinet of weird and wonderful things, bones that no one could identify, and we'd go in and take something out and look at it, argue about it, put it back, take it out again. It was like doing field work right there in the museum. Clive Coy found several Cretaceous bird specimens in that cabinet."

Tim left the Tyrrell to work in the Royal Saskatchewan Museum in Regina, and between trips to the north to dig up things like twenty-four-foot fossil crocodiles, he fed his interest in dinosaurs by visiting the Eastend area whenever he could get away. The Frenchman River valley is so fossiliferous that the geological unit it flows through, called the Frenchman Formation, has become famous for dinosaur material. In August 1991, Tim and his boss, John Storer, and Eastend's public-school principal, Robert Gebhardt, an amateur fossil hunter, were prospecting in the valley south of town when Gebhardt turned over a rock and realized it was a fossilized bone. Tim and Storer took a look and Tim dug farther into the hill with his knife. The bone was a large vertebra, broken in half so that they could see its internal structure, which was hollow and chambered, "like the bones of a bird," Tim said. That meant it had to be theropod. Buried beneath it were some huge teeth. Tim thought they were tyrannosaur teeth, and Storer agreed that it was possible.

Despite the fact that *T. rex* is the most universally recognizable dinosaur in the world, being to dinosaurs roughly what Mickey Mouse is to rodents, not as much is known about it as one would think. For one thing, there are very few complete specimens known: the first was found in 1902 by the American bone hunter Barnum Brown[2] in Hell Creek, Montana. The Eastend *T. rex* is only about 65 per cent complete, but it was only the twelfth to be found, and the second in Canada. If hadrosaurs were coal, *T. rexes* would be diamonds. For another, there is so much variation among the specimens that generalizations are suspect: no two skulls are identical, for example. Tim says the body of his *T. rex*, which he named Scotty because they celebrated its successful extraction with a bottle of Glenfiddich, is more gracile than other *T. rexes*, meaning slenderer, but its skull is "more robust," that is heavier, bigger-boned. "It's sort of like having a pit-bull's head on a greyhound's body." These, he admits, are impressions only, since the skeleton is still mostly encased in the original ironstone matrix.

"We can go back to the research station and I'll show you the specimen," he said, shutting down the computer. "Then we'll drive out to the site."

At the station, we walked through the reception area into a small display room, which contained casts and models of the various animals that had been discovered in the Eastend area. There was a huge *Triceratops* frill and a complete *Thescelosaurus*, a small, primitive relative of the hadrosaurs, whose skeleton seemed to have been held together by a web of tendons, many of which had fossilized so that

2 Barnum Brown was named, appropriately enough, after P. T. Barnum, the circus owner. Brown never missed a chance to capitalize on his discoveries, and used to brag that his fossil finds constituted "the greatest show unearthed."

the specimen looked like a pile of bones covered with stone dowels. We passed through this room into the work area, a large, brightly lit lab dominated at the centre by a huge half-ball of plaster, measuring about six feet across, lying on the floor like a huge, upturned devilled egg. Inside was the *T. rex*.

"Looks like a giant owl pellet, doesn't it?" Tim said glumly as we stood looking down into it. It certainly was the most convoluted specimen I'd ever seen. Paleontologists like to find their dinosaurs lying flat on the ground, stretched out like a patient etherized on a table. This one looked more like what you'd find at the bottom of a stock pot after someone had poured cement into it. Tim and a few volunteers had been picking away at the skeleton for four years, and still had only about half of it exposed. The rock was too hard for dental picks and awls. The only tools that worked were air scribes, like dentists' drills that ran on compressed air, and even those were slow going. They'd made somewhat better progress with the skull, which was in a separate jacket lying on a large table. Above it, on the wall, was a painted *T. rex* as it may have looked in an X-ray, with all its bones shown in outline. The bones that had so far been extracted from the owl pellet were coloured red. There were still a lot of bones in outline.

"We don't know how much of it is in there, exactly," said Tim. "Maybe sixty-five per cent. The jacket weighs six tonnes, I know that. We had to get a flatbed truck into the site and hoist the thing on with a crane."

I was anxious to see the site, so we got into Tim's truck and drove out, heading south from Eastend for about ten miles, then east and south again, and finally turning into a rancher's field on a rutted lane that wound between rounded hills beside a stream. "I really should have blindfolded you," Tim said. "The rancher here doesn't like a lot of people knowing where the site is. There was a lot of traffic when we were taking it out. Hundreds of people a day.

The rancher set up a booth at the site and sold T-shirts, potato chips, Aspirin and film. But now that's all died down and he'd prefer we left it alone."

"I'll cancel the tour bus I hired," I said.

"Don't laugh," said Tim. "We had to organize buses from the Research Station, put up rope barriers, charge admission. We had over seven thousand visitors. Everyone wanted to take home a rock from the quarry."

We were driving along the Frenchman River. The horizon was high above us, cut off by gravelly hilltops covered with sagebrush. The road took us through grassy wetlands fringed with wolf willow. I noticed a rough-legged hawk circling high above the valley, and a Swainson's hawk farther ahead. We were still in carnivore country. Tim pulled over near a spot where a small creek trickled down from between two mounds, and we got out and walked up the stream. Tim pointed to the top of one of the mounds, where a wooden stake protruded a few feet from the peak.

"That stake marks the K/T boundary," he said. The boundary here was a few inches of black rock that contained the extinction of the dinosaurs. Above it the Earth belonged to the mammals. Birds continued to own the sky.

The *T. rex* site was at the bottom of the ravine, opposite and about twenty yards below the stake. It looked as though a giant bite had been taken out of the hillside, or like an excavation for a patio, except I knew that the tons of dirt had been removed a spoonful at a time. The floor was as hard as concrete, but the walls graduated up to a fine sandstone. I got down on my knees and swept some sand off the floor with my hand, and thought I could see more bone.

"Did you get it all out?" I asked.

"No, there's still more down there," Tim sighed. "We'll come back when we get more funding. I guess they figure we've got enough to work on for the time being."

"They" being the Royal Saskatchewan Museum, the parent body of the Eastend Research Station, Tim's employer. A lobby group in Eastend was trying to persuade the province to build a big, modern, architecturally marvellous museum in town to house the *T. rex* and all the other fossils that had been recovered from the area, to turn Eastend, in other words, into another Drumheller. This was the opposite pressure to the kind Rodolfo was under in Plaza Huincul, where he was being urged to downplay the fossils and dust off the sidesaddles. Here there was the usual community split over the issue, with local businesspeople behind the new museum and ranchers and ordinary citizens, who liked their Eastend quiet, simple and manageable, urging caution. Tim seemed to have one foot in each camp. If such a facility were ever built, the *T. rex* from this unlikely gorge would be the centrepiece. Perhaps, as in Jurassic Park, it would be shown fending off attacks from smaller raptors. More likely it would simply greet visitors in the lobby with open jaws, frightening small children so that they would want to come back again and again. However it was presented, it meant that Tim would have to get busy exhuming the specimen in the Research Station, then come back and take out the rest of the bones under our feet, and that was becoming an ever more horrendous task. When it was finished, his life would be changed forever. The whole thing seemed to weigh him down.

WE RETURNED TO EASTEND and Tim went back to his computer. I drove out to visit Sharon Butala, a writer I had met a few years previously and who, in books like *The Perfection of the Morning* and *Coyote's Morning Cry*, shares a deep intimacy with the prairie landscape and its creatures, both above and below ground, living and long, long dead. Sharon's house was just outside Eastend, and as I turned into her lane I saw a marbled godwit rise like a portent from the Frenchman River, which flowed by her house, and settle on the post by the side of her road. I delighted in the bird's ridiculously long, slightly upturned beak, three times the length of its head, and decided that the name "godwit," the origin of which is unknown, must be a homage to God's wit, which had obviously struggled to come up with one more implausible beak shape for this bird. The poor godwit seemed to be sitting on the fencepost showing off, like Jimmy Durante, its outlandish proboscis: "Ha-cha-cha-cha-cha!"

Sharon didn't know quite what to make of the *T. rex* hoopla. She knew that the town had been in decline when the dinosaur was found, but in an article she wrote for a local magazine, she admitted to being somewhat resentful that, in order to visit the site, which was on land that belonged to people she had known since they were preschoolers, she had to "phone to reserve a seat on a bus, drive eighteen kilometres into town, pay twenty dollars, and line up with strangers from California and Ontario." She was especially annoyed when, after all that, the bus passed by her house and the tour guide pointed it out as an added attraction. She didn't actually agree with local fundamentalist Christians who told her they did not believe in the existence of dinosaurs, or else that tiny dinosaurs had been allowed on the Ark but died out when the Flood receded because the vegetation and climate had changed. But she did sympathize, deeply, with the rancher who, at a meeting in which townspeople were informed that the new museum might draw one hundred thousand tourists a summer to Eastend, burst into tears.

The hardest thing to contemplate is change, she told me that evening as we stood on her porch watching the setting sun play over the hills beside her house. Flat country makes for long shadows. The land we were looking at had been given to her husband, Peter, by his father the day Peter turned eighteen, more than forty years ago, and they had never allowed it to change, never ploughed it, hardly ever grazed it. Just kept it, like a faith, a bloodline to the past, "unbroken," Sharon said, putting a rancher's emphasis on the word but also aware of its spiritual connotation. Thinking that dinosaurs had once roamed on it, she said, was scary, as though in looking at what she thought of as Peter's field she was in fact looking at a different planet. Only paleontologists and poets, she had said in the article, could grasp the enormity of *Tyrannosaurus rex*, and there were no paleontologists or poets on Eastend's new tourist board. Later, however, when we went for a walk behind the house, we strolled along a dried creek bed and I bent down to pick up a fragment of bone that caught my eye. I thought it might have been dinosaur, and Sharon was as excited by the prospect as I was. When I showed the fragment to Tim the next day and he told me it was not dinosaur but bison, a mere few hundred years old, Sharon said: "Even better."

In *The Perfection of the Morning*, Sharon describes finding a mule deer skull in one of her fields, and then, as she picked it up, turning to see twenty-two mule deer standing on a hill above her, watching. "I felt blessed in having found so perfect a skull," she writes; "when I turned and saw the deer hovering over me, watching me as I picked it up, admired it, holding it with reverence, as if it were somehow holy, it seemed to me as if they had given me the skull, as if it were a gift to me." Perhaps, because there are still living bison, the bison bone was a gift in a way that it could not have been if it were dinosaur, a connection with a living past. But if birds are dinosaurs, if she could pick up a dinosaur bone and then, looking up into the sky, see a rough-legged hawk riding the thermals above her, she might feel the same sense of privilege, feel herself, as she had with the deer skull, as I had with dinosaur bones, "moving close to a level of understanding about the nature of existence."

THE NEXT DAY, AS I was helping Tim unpack books from the boxes in his bookstore, I found a copy of William Beebe's *Jungle*

Peace. I'd been interested in Beebe since reading his 1905 book, *The Bird: Its Form and Function*, which had included his photograph of the ostrich's claws and an illustration of what he thought a *Hesperornis* might have looked like. *Jungle Peace* was about his attempt, in 1917, to establish a bird research station deep in the Venezuelan jungle. I read it by candlelight that night in the van, which Tim let me park in his driveway. When Phil had suggested that the ostrich figured in the link between dinosaurs and birds, he had also mentioned a strange creature called the hoatzin, a South American bird with significantly aberrant dinosaur affinities. Two of Beebe's chapters dealt with a side trip he made into British Guyana to look for hoatzins. Though members of the Galliforme family and therefore related to pheasants, grouse and chickens, hoatzins are living fossils, or what T. H. Huxley called "persistent types," organisms that have remained essentially unchanged over enormous periods of time.[3] Evolutionary lacunae; animals forgotten by time, very attractive to the Victorian mind. Young hoatzins, when still flightless, retain two reptilian clawed fingers at the tips of their wings. Beebe found a colony of them on the Berbice River, in a district that was then called New Amsterdam, and described a featherless nestling peering at him from its deep, woven nest above the river: "Higher and higher rose his head, supported on a neck of extraordinary length and thinness. No more than this was needed to mark his absurd resemblance to some strange, extinct reptile. A young dinosaur must have looked much like this. . . ."

When one of Beebe's assistants climbed out onto the limb to catch the young hoatzin, the nestling "stood erect for an instant, and then both wings of the little bird were stretched straight back,

---

3 Huxley classified the hoatzin as the type and sole member of the family Heteromorphae, the mixed forms, displaying characteristics of both birds and reptiles. "In view of the immense diversity of known animal and vegetable forms," he wrote in 1862, "and the enormous length of time indicated by the accumulation of fossiliferous strata, the only circumstances to be wondered at is not that the changes of life have been so great, but that they have been so small."

not folded, bird-wise, but dangling loosely and reaching well beyond the body." Beebe was quite aware that he was witnessing a bird regressing to the dinosaurian state: "For a considerable fraction of time he leaned forward. Then, without effort, without apparent leap or jump, he dived straight downward, as beautifully as a seal, direct as a plummet and very swiftly." The nestling remained underwater for nearly five minutes before its reptilian head poked above the surface. It dove again, and six minutes later reappeared closer to the edge of the pond, where some pimpler branches hung just above the water. There, "a skinny, crooked, two-fingered mitten of an arm reared upward out of the muddy flood and the nestling, black and glistening, hauled itself out of water. Thus," marvels Beebe, "must the first amphibian have climbed into the thin air."

Long before he got to the hoatzins, however, I was caught in the intricate weave of Beebe's writing, for he was as interested in people and the places where they lived and the language they spoke as he was in birds. *Jungle Peace* is much more about travel and adventure than about science. When his ship stopped in Martinique, he found the island "filled with a subdued hum" of human voices, "a communal tongue, lacking individual words, accent and grammar, and yet containing the essence of a hundred little arguments, soliloquies, pleadings, offers and refusals." On Barbados, during an eclipse of the sun, he observed how, when "walking beneath the shade of dense tropical foliage, the hosts of specks of sunlight sifting through, reflected on the white limestone, were in reality thousands of tiny representations of the sun's disk incised with the segment of the silhouetted moon, but reversed, like the image through the aperture of a pinhole camera."

In Berbice, a quiet town deep in the interior of New Amsterdam, at the end of the Pomeroon Trail, he spent his days wading through red tape in Colony House and his nights at "the club, the usual colonial institution where one may play bridge or billiards, drink swizzles, or read war telegrams delayed in transit." One night after dinner the steward approached him diffidently:

"Would the sahib like to see the library?"

He was led up flights of protesting stairs to an upper room, "barnlike in its vacantness," its walls lined with books. "There was an atmosphere about the room which took hold of me at once . . . something subtle, something which had to discover itself." As he nosed his way along the dustless shelves, each volume aligned with perfect precision, "the secret of the place came to me: it was a library of the past, a dead library." No one used it, no books had been added to it for years, "most of them were old, old tomes richly bound in leather and tree calf." Little-known histories and charmingly naïve reminiscences: *Lives of the Lindsays* and *The Colloquies of Edward Osborne, Citizen and Clothmaker of London.* Thier's *The Consulate and the Empire* and the *Memoirs of the Lady Hester Stanhope, as Related by Herself in Conversations with her Physician.* There were also first editions of Dickens and Scott and Baron Cuvier, whose adamant rejection of Darwin's concept of natural selection, Beebe realized, the hoatzins had so completely refuted.

After selecting a book to take back to his room, he turned out the light, then paused for a moment to look about. "The platinum wires still glowed dully, and weak moonlight now filled the room with a silver greyness." He wondered whether, "in the magic of some of these tropical nights, when the last ball had been pocketed and the last swizzle drunk belowstairs, some of the book-lovers of olden times, who had read these volumes and turned down the creased pages, did not return and again laugh and cry over them. Such gentlefolk as came," he imagined, "could have sat there and listened to the crickets and the occasional cry of a distant heron, and have been untroubled by the consciousness of any passage of time." The library at the end of the Pomeroon Trail, Beebe's own lost world.

TIM'S OWN LIBRARY, which I perused in his house the next morning over several cups of coffee while Tim was out walking the dogs, was a wonderful blend of science and literature, and included many books by and about Darwin, Huxley and the huge, glowering lump of evolution that sat in the Victorian stomach like too much plum pudding after Christmas dinner. Early English mariners called the hoatzin "the Snake-eater of America," wrongly, as it turned out, because it is vegetarian; its chief food is pimpler leaves. It has an extremely complicated and bulky digestive system to deal with all that roughage. In 1867, Huxley delivered a series of twenty-four lectures to students at the Royal College of Mines (one of whom might have been George Mercer Dawson) in which he pointed out that, just as some birds had dinosaur-like digestive systems, so some dinosaurs had bird-like hearts, pneumatic lungs and probably "hot blood," which meant the two groups were related, i.e., from the same province, but not necessarily that the one was descended from the other. Not until he read Ernst Haeckel's *Generelle Morphologie der Organismen*, in which Haeckel showed how the embryos of a species developed through all the stages that the species had evolved through, how a human embryo, for example, looked first like a fish, then like a frog, then like a baboon and finally like a human, did he begin to think in terms of bloodlines, and consider how a living animal could be descended from a fossil, how *Archaeopteryx*, for example, could not only be from the same province as the hoatzin, but could actually be ancestral to it. Haeckel introduced the element of time into the equation.

The recurring irony for Darwin was that, as he stated in *The Origin of Species*, the truth of his hypothesis stood or fell by the fossil record, and the fossil record kept turning up new specimens that both confirmed and denied evolution. Although the discovery of *Archaeopteryx* seemed to confirm it by providing a handy missing link between reptiles and birds, the discovery of homonid remains in the

German valley of Neander tended to cast serious doubt on the smooth line of human evolution. Darwin, as understood by Huxley, or rather as interpreted by Huxley, had suggested that evolution progressed in such a way as to create ever better adapted members of a species to its environment. In prolonged dry spells, the lungfish that could somehow survive out of water got to become an amphibian. In human beings, that meant (to Huxley and others) ever increasing intelligence. Biologically, that meant ever larger brains. Neanderthals, however, had bigger brains than modern *Homo sapiens*. This was a problem; they couldn't have been better adapted than us, because they died out. Many Victorians, including some eminent scientists, solved the dilemma by saying that evolution did not apply to human beings, but, after reading Haeckel and having rejecting God, Huxley knew that it had to. As Loren Eiseley observes in his book *Darwin's Century*, "so long as man was regarded as 'outside' of nature, a unique being divorced from any but the most recent past, he stood as a challenge to all scientific attempts to explain, not alone his own origins, but those of even the 'natural' world about him."

BEFORE LEAVING EASTEND, I visited the local museum because I wanted to see what might be lost if the government built a spanking new, dust-free dinosaur facility on the edge of town. The museum was on the main street, just down from Jack's Café, where Tim and I had lunch. Over our burgers and beer, Tim told me that the museum contained fossils collected in the 1920s by the late Harold "Corky" Jones, Eastend's resident eccentric.

Paleontology and ornithology have something else in common besides birds, and that is the valuable contributions that have been made to them over the years by knowledgeable amateurs.[4] Corky

---

4 Many scientists acknowledge this debt. Jack Horner, in *Dinosaur Lives*, writes that "the weekend collector and the seasoned paleontologist form a natural alliance, because their ambitions are the same: to experience the joy of discovery, thereby increasing what we know about the world in which we live."

Jones was born on the Isle of Wight in 1880, where his father, a doctor who attended Queen Victoria when she visited the island, took him fossil hunting along the island's slaty coast, I imagine much in the way that Charles Smithson, in Fowles's *The French Lieutenant's Woman*, collected trilobites during rambles on the rocky shores of southern England. Jones had been heading west for the Klondike Gold Rush in 1898 when he fell in love with the Prairies and got off the train at Maple Creek, a few miles north of Eastend. He worked on a ranch for a while, married the organist at the Maple Creek Anglican Church in 1911, and then drifted down to Eastend in 1918, where he decided he was a mechanic and opened a garage. In 1930, he became the town's sole constable. He liked policing, he said, because all he had to do was ring the town's curfew bell every night at nine o'clock, which left him lots of time to indulge his "mania for fossils." As a lawman, he seemed to have been somewhat easygoing. To control traffic at the main crossroads, he set a wooden post in the middle of the intersection and had automobiles go around it. The post became known as Corky's Constabulary. One morning someone came into his office and told him the post was leaning over at a dangerous angle, and Corky went out and arrested it for being drunk on duty. The American writer Wallace Stegner, who grew up in Eastend,[5] and who wrote about the experience in his most famous book, *Wolf Willow*, changed the names of all the townspeople except for Jones's. A few years ago, when Tim asked him why, Stegner replied: "Because Corky was the only one I ever liked."

When Charles M. Sternberg, the well-known bonehunter who collected for the Royal Ontario Museum and was then working for the Geological Survey of Canada, travelled through Eastend in 1921, he dropped into the town office and happened to notice that

---

5 Then called Whitemud, another descriptive name, and one that distinguishes it from the Mississippi, which Borges somewhere refers to as "the mulatto-hued river."

the clerk was keeping his door open with a *Triceratops* horncore. It was Corky's first find. Sternberg and Corky maintained a close and lifelong friendship. For years, Corky would find things and send them to Sternberg, who would identify them and send them back, and Corky would store them in the school basement. In 1932, for example, Sternberg wrote to Corky: "I am returning today, under separate cover, the bone you sent me for identification. . . . It is an ungual phalanx or claw bone of a large carnivorous dinosaur, *Tyrannosaurus rex*." A year later, in June, Corky reported the end of a prolonged drought: "It is a changed country since my last letter," he wrote. "There is more grass and the country looks greener than it has done for years. Several things in the garden that we had thought dead are coming to life. It's as if some wizard had waved his wand over the land." Over the years, Corky collected two more important fossils: a complete *Triceratops* skull and the neck fringe of a *Torosaurus*. Both these and other treasures, Tim said, were on display at the Eastend museum.

The building that housed the museum was built as a theatre in 1917 by Richard Haddad, who sold it to Bob Dane, who sold it to Lyle Watson, who donated the oak display case that contained photographs of most of the town's principal buildings as well as of the 1952 flood that swept down Eastend's main street, stranding people in their cars, inundating the basement of the school and destroying a great part of Corky's fossil collection. The photograph looked a lot like the photograph in Rodolfo's museum, depicting the Plaza Huincul flood of 1954. A second large case contained a "Dutch Display," and included a pair of wooden shoes once worn by Georgette Anderson's mother. I loved these items for the glimpses they provided into the real lives of people who settled in the area; it was like walking down a quiet street and gazing into softly lit windows. There was Mrs. Crawford's parlour organ, and Dr. Toth's dentist's chair. On a headless dressmaker's dummy was the black crepe dress worn by Sarah Bohen to her great aunt's funeral, May 12, 1911. Over here, under thick glass, an old camera, three straight razors, a shaving brush, some chipped china plates and an assortment of coal-oil lamps. All these precious items were identified and described on small, white index cards typed on an ancient typewriter whose "e" and "a" had not been cleaned in some time. And, of course, there was dust everywhere, especially in the science corner, where what was left of Corky's work was laid out for the world to admire. The famous *Triceratops* skull and a cast of the *Torosaurus* frill, a huge, scalloped and perforated neck piece such as might have been worn by some gigantic Victorian monarch. Beside these, pinned to the wall or laid out along a narrow bench, were *Triceratops* shoulder blades and vertebrae, several *Trachodon* tail bones, the foot of an *Ornithomimus*, and a Cretaceous turtle shell (Basilemys), the humerus, jawbone and horncore of a Titanothere ("related to rhinos and horses"), and, on the proscenium of the old stage, a complete *Brontothere* skeleton, larger than any bison. From these I moved on to a photograph of Eastend's first automobile, a McLaughlin, owned by Mr. J. C. Strong (undated), the skull of a grizzly bear, collected locally, several ammonites and fossil sponges, an ice-age bison horncore, a mammoth tooth and the *pièce de résistance*, Corky's field box, still containing the homemade tools he used to collect fossils. Touching in its simplicity, scary in its familiarity, the box said much about the conditions under which Corky worked and the love he bore for worlds we can still only barely comprehend, for preserved in a battered La Palina cigar box ("2 for 17 cents") were a rusted geology hammer, a few blunted drill bits of various bores, a length of flattened copper pipe, two worn paint brushes and the flat, iron tongue from a door latch.

"It is a question," Stegner wrote in *Wolf Willow*, describing this very collection, "whether or not the museum means anything outside of an easily satisfied and idle curiosity. . . . It offered [schoolchildren] more information on the history of their own place than anything else.... And even if no Whitemud child takes fire from Corky's collection and becomes an anthropologist or paleontologist, something may still have been accomplished by his example. Any child who knows Corky can see knowledge being loved for its own sake."[6]

AS I DROVE OUT OF Eastend, I glanced at my gas gauge and saw that I was nearly out of gas, but I thought I had enough to get to Maple Creek. I forgot that I was still travelling uphill, and the needle bottomed out after a surprisingly few kilometres. I sputtered into Robsart to find a gas station, but all I found were empty, clapboard houses and a dust-filled general store, farm supplies still stacked in the window and obviously undisturbed for many years. I sputtered back out to the main highway and continued climbing towards Maple Creek. The incline was so steep that I could not get the van to go faster than sixty kilometres an hour. I found myself urging it on with short, forward thrusts of my upper body. I considered cutting cross-country, as the land there was hilly but apparently unobstructed by fences or coulees. I thought about Barry Lopez who, while driving across a desert, put his Jeep in low gear and got out to walk beside it, for a break, like a man with a dogsled crossing the Arctic barrens. Easier on the dogs, too. For a while I considered trying that.

Then, just as the van topped a small rise and was slowing down to attack another small rise, my windshield suddenly darkened. For a moment I thought I had gone to sleep, but then something like reason returned and I realized that I was looking at the underside of a large bird, a hawk. It had started up from the side of the road, startled by my approach, and had very nearly flown into my

---

6 This quotation and the two from Corky's Sternberg letters are borrowed from Tim's article, "H.S. 'Corky' Jones: A Biographical Note, " published in *The Blue Jay* in September 1986. Tim never met Corky, but runs into him every day in one way or another. As every paleontologist knows, time is rock, built up layer upon layer, each individual stratum being indecipherable and therefore meaningless without the strata immediately above and below it.

windshield. Its wingspan filled my vision entirely. Grasped in its talons was a large, writhing snake. For half a heartbeat all I could see was a flurry of feathers, a mottled, downy chest, the stern, hawkish expression, and a length of squirming muscle in its claws. Then it was gone. The entire surreal episode had lasted only a split second, like a subliminal image of the Mexican flag inserted between frames of a movie about the Cypress Hills. Had I hit it? Had it even happened?

I had not hit it, at least, for I could see it settling placidly on a fencepost ahead of me, snake still wriggling beneath it, albeit with a certain air of hopelessness. It was not, I was tempted to think, a particularly intelligent hawk, having chased the snake onto the highway like a toddler chasing after a ball. But perhaps it had not evolved to deal with Volkswagen vans, or perhaps evolution had given it a different kind of intelligence.

When Dale Russell realized how intelligent some of the smaller theropods had become by the end of the Cretaceous Period and tried to imagine how smart they would be today if they had not disappeared, if they had had another sixty-five million years of evolution, he began his thought experiment with a *Troodon*. He worked out its encephalization rate, calculated which organs would recede and which would assume more prominence, what bifocal vision would do to its facial structure, what bipedal stature would do to its tail, and came up with something that looked vaguely humanoid; upright posture, striding gait, bright, intelligent eyes, wide mouth and large, hairless, domed head. As I mentioned earlier, the result was decidedly anthropomorphic, a creature that looked like an amphibious aluminum-siding salesman. This was a logical step for Dale, who is an extremely religious man, a devotee of Pierre Teilhard de Chardin, the paleontologist priest who believed that the goal of evolution was to produce human beings who could be fully conscious of the existence of God.

But now it occurred to me that there was another way to approach the evolution of dinosaur intelligence, and one that would not necessarily lead to the conclusion that the more intelligent an organism became the more it began to resemble human beings. You could look at how smart birds have become.

True, most birds, chickens for example, appear to be only as smart as they need to be, and chickens don't need to be very smart. It hardly seems likely that a lot of abstract thought is going on behind their worried faces as they strut about waiting for something to throw them a handful of grain. Hawks spend a lot of time sitting on fenceposts waiting for something to move in the grass, an occupation that does not on the face of it require an advanced degree in metaphysics. Owls, their nocturnal cousins, have a reputation for wisdom that is difficult to explain. About the smartest thing a burrowing owl does is twist its head in a complete circle if you walk around it as it stands beside its burrow. Cowboys used to do that for sport: Will James, the cowboy-artist-novelist of the 1920s who broncoed for a spell in southern Saskatchewan, wrote about cowboys riding their horses around and around burrowing owls, just for the goshdarn fun of it. One doubts that, as the owl pivoted its sagacious head, it was thinking: "Cowboy, know thyself."

If we call intelligence the ability to entertain abstract thought, then most birds have low IQs. But, as I've said, we're the ones who design the IQ tests. Abstract thinking is a human measure of intelligence, and bird intelligence may be different from human intelligence. We watch a bird building a nest and think, since it is merely reproducing some geonomic imprint of the perfect nest, it is behaving no more intelligently than when it incubates an egg or stirs up crayfish. Scientists say such instinctive behaviour is hard-wired. We see a bird's tiny brain and think: birdbrain. In fact, birds must think differently from us because they think with a different part of their brains. The functioning part of the mammalian brain (by which I mean the part of the brain the function of which we think we understand) is the cerebral cortex, but birds do their thinking in a region called the hyperstriatum. Mammals don't even have a hyperstriatum. We can't really know what form thoughts take in such an organ. Even contemplating it makes us giddy, as though birds are really some alien life form beamed to our planet to study or possibly to help us.

But there's a simpler explanation. Birds, being descended from dinosaurs, have been evolving intelligence in an unbroken line for 200 million years. Wouldn't it make sense if they thought in different ways from us? Isn't it logical that, as the cowboy circles the owl, something different, something more complex, something infinitely unknowable to us, is going on in the owl's hyperstriatum, some process taking place that the cortex-limited cowboy cannot even guess at?

There must be other forms of sentience than ours. I have walked through the woods near our cabin in winter and been swarmed by a small flock of chickadees fluttering around my face, pausing for seconds on branches inches from my eyes, silently assessing this unaccustomed life form moving among them. I stood still, and within minutes chickadees alighted on my shoulders and peered into my ears. I held out my hand and chickadees gripped my fingers. My heart hardly dared beat. I did not receive the impression that I was being mistaken for a tree, that my fingers were being taken for twigs. I felt I was being reconnoitred, sampled, assayed and, inevitably, found wanting.

CHAPTER ELEVEN

Dry Island

IN SPITE OF EVERYTHING, I made it to Maple Creek without running out of gas, and rejoined the TransCanada suitably tanked up for the final sprint to Alberta. I first noticed the storm clouds in my side mirrors as I was crossing the border. Behind me, the blue Saskatchewan sky had darkened and swelled like a vast, cosmic bruise, and when I looked ahead, the sky there, too, had suddenly filled with black clouds so heavy with water they seemed to be resting on the ground, the way someone might put a pail down for a moment to catch their breath. I was, apparently, still climbing, and now, moreover, was driving into a roaring headwind, hardly making any headway at all. I kept pulling over to let swifter cars and trucks, the Pintos and the Broncos, pass me, which they did with little surges of joy, like pet horses let unexpectedly off their leads.

In this fashion I barely reached Brooks by nightfall, and there turned south to Kinwood Island Provincial Park. Being on the Prairies, the word Island in the park's name didn't immediately register with me, but it turned out that the park was, indeed, on an island in the centre of a large body of water called Lake Newell, the largest man-made reservoir in Alberta. If I had continued on for another twenty kilometres or so, I would have come to Bassano, where a dam on the Bow River allowed the town to advertise itself as "The Best in the West by a Damsite." What Bassano might be equally proud of was the fact that it shared a name with an ancient body of water known to geologists as Glacial Lake Bassano, an accretion of meltwater that built up over thousands of years behind a huge glacier during the last Ice Age. Glacial Lake Bassano was so immense that when the Ice Age ended, rather abruptly, it seems, and the glacier dam melted and burst, the waters that rushed out carved, in a matter of hours, what is now the Red Deer and South Saskatchewan river systems, scooping out so much soil as they went that they created the Badlands right down to their Cretaceous sandstone bases. It was thanks to the water released from Glacial Lake Bassano, in other words, that so much dinosaur-bearing rock was exposed. Global warming isn't always a bad thing, in the long run.

When I stopped at the park entrance to be given a campsite, beyond the kiosk I could see swaying trees with very respectable trunks, some of them being blown horizontal by the wind, their crowns actually splashing maniacally in the lake, collecting foam in their vibrant leaves from the whitecaps on the waves. Gulls whizzed by helplessly overhead, pale snatches in the darkness, who knew where they would end up? Montana, maybe. Rain throbbed against the side of the van as from a fire hose.

From a pay phone, I called Paul Johnston in Drumheller. I would have called Phil, but I thought he and Eva would be out at the *Albertosaurus* quarry in Dry Island Buffalo Jump Provincial Park, where I was supposed to join them. When I told Paul I'd be a day or so late, he said not to worry, the rain up there was so torrential the field team had slunk into Drumheller for a few days to dry off. The weather was so bad, he said, that he and his wife, Noreen, had had to move into the house.

"Move into the house from where?" I asked.

"From our teepee out back," he said, explaining that when their kids were teenagers, he and his wife, Maureen, bought a big canvas teepee, set it up in the backyard and moved into it. They spent whole summers out there, cooking on a central fire, sleeping on a futon mattress in sleeping bags, hardly having to go into the house at all except to use the bathroom. It was like a permanent field camp. They got so used to it that when they noticed that the kids had grown up and left home, they just stayed in the teepee. But this week the rain had forced them inside.

"You might as well stay where you are and just go directly up to Dry Island in a day or two," Paul told me.

It was too cold and wet and noisy to sleep, so I sat up much of the night reading. In the morning, the wind had not abated at all and, although it was supposed to be summer, the temperature continued to drop. It was June, and there was talk of snow in Calgary. Eva had given me precise instructions for finding the field camp at Dry Island, and as I drove north I was greeted by a warm sun and a following wind. The van stepped lively, and I made good time, arriving at the camp shortly after lunch.

It was located at the end of a long lane that led past a farmhouse and through a rusted Paige-wire gate into an open field. There were a few tents in the field, but no one was in sight. A large trailer stood on blocks to the right, with a canvas-covered structure beside it, a kind of square version of Paul's tepee that, when I looked inside, turned out to be a preparation tent: the floor was littered with small plaster jackets containing bones from the quarry waiting to be more finely prepared on rainy days. A tarp stretched from the side of the trailer over a long row of wooden picnic benches. Feeling a bit like Goldilocks, I manoeuvred the van into a sheltered spot beside some trees, then walked over to the trailer and looked inside. It was a real kitchen, with a propane stove at one end of a counter and refrigerator at the other, a stainless-steel sink between them and shelves of plates and cutlery and dry food in neat boxes. A rack of kitchen knives hung on the wall above the stove. One whole cupboard was devoted to vials of herbs and spices. The floor was clean. Bowls of snacks had been set out on the table.

Munching a handful of trail mix, I strolled across the field to the edge and peered over a sharp drop into the deepest Badlands I had ever seen, eight hundred feet straight down. Flat-topped mesas rose from the valley floor in the distance, shimmering as in some weird, futuristic painting, a cover illustration for a science-fiction novel. Here was Thomas Browne's depiction of the bottom of the Mediterranean Sea, the water all drained out of it. The view made me feel light-headed, as though I were standing at the edge of the known universe, looking down into what most of Europe thought Columbus would see instead of America.

"You can't see Dry Island from here," a voice called from behind me. I turned and saw a woman approaching from behind the kitchen trailer. "I'm Martha," she said. "I'm the cook."

Just then two men walked into the field from farther along the lip of the Badlands. One of them was Jørn, whom I had last seen a month before at the bus station in Plaza Huincul, steeling himself for a nineteen-hour bus ride to Santiago. With him was one of the Tyrrell's field camp leaders, who introduced himself as Stewart Wright. Both looked as though they had been lolling in mud. I had seen Jørn like that before, in Argentina; in fact, I had rarely seen him looking any different.

"We've just been scraping off the quarry," he said. "I thought we had rain in Patagonia, but I've never seen anything like the rain we had here these past few days. And wind. My tent blew down, Phil and Eva gave up and went into Drumheller, the quarry was filled in with a mudslide. What a mess. Did you by any chance bring some Breeders' Choice?"

Alas, I hadn't. I told him about the forty-ouncer I'd bought before leaving Plaza Huincul. I'd packed it in my suitcase wrapped in a four-hundred-dollar suede jacket, and of course it had shattered during the return flight. The jacket had been ruined and several books soaked, including my bird diary.

"Too bad about the whisky," said Jørn, sympathetically.

"I'm afraid all I've got is a bottle of Glenfiddich," I said, and he perked up.

Martha said, "I'll make the coffee," and before long the four of us were sitting at one of the picnic tables, sipping a version of Jørn's elixir and feeling better. It was good to see Jørn again; he was one of those stalwart types who make any situation appear tolerable. In Patagonia he had come down with some kind of flu and had spent the last few days slumped in the corner of the quarry, propped against the cliff wall in the rain, burning with fever and making jokes about the food. Stewart was from British Columbia. He had long, black hair and wore wraparound sunglasses and a black base-ball cap turned visor-backwards, a black T-shirt and black jeans. Five years ago he'd been putting a sort of life together in Nanaimo, on Vancouver Island, when he'd been hit by a bus and lost the use of his left arm. The doctors told him it was a permanent condition, but he started exercising and eventually was able to hold a hammer. "I realized that if you just sit around waiting for things to happen to you, there's a pretty good chance that what happens to you will be bad. I had always wanted to be a paleo, I don't know why, but ever since high school I've loved being outdoors, and I've loved dinosaurs. So when I got some money together I signed up for one of the Tyrrell's field programmes, and worked my ass off all summer. I thought my arm was going to fall off, but it didn't, it got stronger, and I came back the next summer, and the summer after that the Tyrrell hired me as a fossil preparator. I'm the happiest guy in Canada," he said.

More people arrived throughout the afternoon, the paying volunteers of the Tyrrell's Field Experience Program. So many people want to help the Tyrrell staff members during the summer that a few years earlier the museum had decided to charge them eight hundred dollars a week for the privilege: each summer, people drive, fly or hitchhike to Drumheller, and are then transported out to the museum's various dig sites, where they are shown how to prospect, collect and prepare fossils. It's an enormously successful programme, and always booked solid. By dinnertime there were ten of us in camp. John A. and John H. had driven all the way down from Yellowknife: in real life, John H. was a mining engineer and John A. was a waiter. Brooke was a graduate student from Alaska, thinking of doing her master's on fossil mammals, and Andy was going into first year university in Iowa, majoring in anthropology. Al was a high-school science teacher from Hawaii, and Ruth was a grandmother from northern Michigan, recently widowed, who wanted to do something different. The two Johns had been here last year, and this was Al's fifth straight season. We were a raggle-taggle group of amateurs who had come to one of the richest bone fields in the world.

After dinner, we piled into one of the museum vehicles and Stewart drove us across the prairie to the head of the trail that led into the Badlands. Tomorrow morning we would walk from here to the quarry, a distance of two kilometres as the crow flew but which, with all the winding and climbing down coulees and over tors, would take us forty-five minutes. It was dusk, and nighthawks were making figure eights above the fields, catching grasshoppers on the down swings. We stood on the cliff edge looking out over an alien landscape as Stewart pointed out the various levels and formations illustrated in differently coloured rock, the circumeroded buttes of ochre and red that were slowly greying in the twilight. Stewart pointed to one that had a wealth of shark teeth. Another had yielded a crocodile skeleton. It was still Late Cretaceous, but a step less recent than the Frenchman Formation in Eastend. I was travelling backwards in time. Right at the centre of our view, rising from the valley floor so high that its flat top was at the same level as we were, was the narrow tube of land known as Dry Island. When the furious waters of Glacial Lake Bassano tore past here, gouging out the Badlands, the top of Dry Island stayed dry, a tall, round pillar completely surrounded by swirling chaos: any animals caught on it would have been stranded there. Now the occasional mule deer might climb up to it, but mostly it was pristine prairie, never grazed by cattle or even by buffalo. We were looking at a time capsule, isolated and untouched for more than ten thousand years. Except by birds, of course. It wasn't nearly as big as Conan Doyle's Maple White Land, but it enjoyed the decided advantage of actually existing.

"What kind of matrix is the *Albertosaurus* in?" Brooke asked Stewart.

"The quarry is actually two quarries," Stewart said. "Half of it is in soft sandstone, and half is in very hard ironstone. We call one side the Beach and the other the Hardrock Café."

"How did it happen that two kinds of rock formed in so small an area?"

Stewart shrugged. "Everyone has a theory about that," he said. "Everyone has a theory about everything. Extinction, herd behaviour, food . . ."

"Bird origins," I ventured.

"No!" said Stewart, somewhat violently, I thought. "That debate's over, finito, kaput. Give it up, Larry Martin! We've got feathered dinosaurs, man. Case closed. You lose!" He stomped back to the truck and started the engine: "Everybody in who isn't walking," he said angrily. I got in.

THAT SPRING, WHILE WE were in Patagonia, Larry Martin, still a stalwart critic of the bird-dinosaur hypothesis, had published in *The Sciences* a long review of three books dealing with the debate, including *The Origin and Evolution of Birds*, by his fellow skeptic Alan Feduccia.[1] It was a skilful bit of rhetoric. He himself, he wrote sadly, had once been a believer, but when he began to look carefully at the claims of the theory's new defenders and saw that most of the supposed shared characteristics between dinosaurs and birds were either misinterpretations or were also shared by other animals, "when the burden of ad hoc repairs became too heavy for me," he wrote, "I had to abandon the theory." He damned Sankar Chatterjee with faint praise for his theory that bird evolution began in the Triassic Period, before the appearance of *Protoavis*, his own earliest-bird candidate, because if *Protoavis* were a bird then it weakened the claims of John Ostrom and Phil that bird evolution occurred much later, near the end of the Jurassic. No one except Chatterjee, not even the other dinosaur-bird believers, seriously thought that *Protoavis* was a bird, so by giving credence to Chatterjee, Martin was making the opposing camp seem disloyal to its own. Martin also stated quite forcibly that the feathered dinosaur from China, *Sinosauropteryx*, was not, in fact, feathered, but probably resembled the modern sea-snake, which has "frayed collagen fibers under the skin [that] can look feathery." He neglects to add that none of the other characteristics common to both *Sinosauropteryx* and *Archaeopteryx* are found in sea-snakes.

"Skepticism is a reasonable stance in any scientific investigation," Martin sighed. "But it is virtually indispensable when

---

1 The other two were *The Rise of Birds: 225 Million Years of Evolution*, by Sankar Chatterjee, and *Taking Wing: Archaeopteryx and the Evolution of Bird Flight*, by Pat Shipman. In 1983, Chatterjee found a 225-million-year-old fossil, which he named *Protoavis*, that he claimed was a bird. Shipman examines the two theories of how animals took to the air, the "ground up" and the "trees down" theories. Proponents of the dinosaur-bird link argue that flight evolved when feathered dinosaurs flew up from the ground; opponents hold that tree-dwelling thecodonts started gliding first, like flying squirrels, and later learned to flap. She ends up siding with the dinosaur-bird side of the debate. Chatterjee's *Protoavis* has not been accepted as a bird, even by the dinosaur-bird proponents; Jacques Gauthier, for example, goes so far as to lump it in with British physicist Fred Hoyle's claim that *Archaeopteryx* was a hoax because evolution, according to Hoyle, was driven by germs from outer space occasionally falling to Earth and infecting our DNA. Phil simply says there's not enough *Protoavis* material from which to make a definite claim one way or another.

confronting a juggernaut, a theory you want so much to believe that it casts a powerful spell over your own good sense."

THE BRITISH MUSEUM of natural history paid four hundred pounds for an *Archaeopteryx* specimen in 1862 and placed it, not among the dinosaurs, but in the Bird Gallery. It was labelled "the Adam among birds." But its long tail and three clawed fingers were clearly dinosaurian, and when John Evans, a scientist friend of Huxley's, examined it, he noted that it also had teeth and a theropod's braincase. This, Huxley maintained, clearly established the link between dinosaurs and birds. When even this hard evidence failed to impress the skeptics, however, Evans wrote to Huxley in exasperation: We have the fossil beak, the fossil teeth, the fossil fingers and the fossil feathers, he said. What more do they want—the fossil song?

PHIL AND EVA ARRIVED early the next morning with Seven, Phil's black and white Shetland sheepdog. Next came Darren Tanke, another Tyrrell technician, who drove up in a battered, blue, 1946 Ford flatbed truck, the tires half flattened by the weight of a thousand-litre water tank resting on its bed. The camp was now complete. Darren looked a bit like a young Teddy Roosevelt, wearing a wide-brimmed felt hat, wire-rimmed glasses and a blue polka-dot bandana around his neck. He climbed out of the truck, and when the introductions had been made I asked him where the water came from.

"There's a Hutterite colony just down the road a ways," he said. "I arranged with them last year to let us use their well. They're very interesting people," he added, "not nearly as dogmatic as you might think. They held a meeting to discuss whether or not to let us have the water. One of them asked me what we were digging here, and when I told him, 'Dinosaurs,' he said, 'Bah! There were no such things as dinosaurs.' I thought, there goes the water, but some of the other men disagreed with him, and a real discussion ensued. I invited them all down to the quarry to have a look, and eventually they agreed to give us the water."

Darren climbed into one of the museum vehicles, Stewart started the second one, and soon all thirteen of us were on our way to work.

We parked the vehicles at the edge of the Badlands and began the long descent into the Cretaceous Period. I was excited, filled with adventure, as though we were retracing Professor Challenger's steps into the Lost World. We followed mule-deer trails for the most part, narrow paths carved into cliff faces and angled over stretches of what looked to be solid, buff-coloured sandstone but which turned out, after the rains, to be of slick, gelatin-like dendritic clay that took us along the verges of steep precipices. Mule deer appeared not to be disturbed by great heights or sloppy footing, or perhaps they considered peering past their feet into a crevasse or canyon several hundred feet straight down to be an effective deterrent to coyotes.

Phil took the lead. He seemed, with his long legs, to stride along as nonchalantly as though he were crossing an empty parking lot. More nonchalantly than that, in fact, since he was far more at home here in the Badlands than he ever would be in a place that had parking lots, even empty ones. A few years ago he had been offered the top scientist's position at the Royal Ontario Museum but had turned it down because he didn't want to live in a city the size of Toronto again, and, as he told me at the time, "ROM didn't have the Badlands in its backyard, as I do here." He spends as much time in the field as he can manage, more than even the Tyrrell Museum is comfortable with, and he has voluntarily reduced his salary by twenty per cent so that the museum's directors have less to complain about when he is absent. He's driven to hunt for bones as few others have been. He worries about the amount of flying he has to do to get to the bonebeds, which is especially difficult when he works in Inner Mongolia. His father died of leukemia at the age of sixty-two, and for most of his life Phil has harboured a superstitious dread that he, too, will not live much longer than that. Now that he is fifty, he says he feels almost as though he is living on borrowed time, that every field trip might be his last.

I had a friend a few years ago who knew he was dying of cancer. All his life he had been a prolific and promiscuous reader, but when he learned he had less than a year to live he became very selective. "Every time a new book comes out I ask myself if it's good enough to be one of the hundred or so last books I will ever read in my life," he told me. "Very few measure up." Perhaps something like that was behind Phil's decision to focus his research so exclusively on theropods and their link to birds. If so, I thought, there must be more significance to this *Albertosaurus* quarry than was immediately apparent.

I soon learned that there was. Finding the first *Albertosaurus* was Joseph Burr Tyrrell's only contribution to paleontology, but it was big enough to justify naming a museum after him. The skull and legbones he dug out of a gravel coulee in 1884 were the first dinosaur fossils to be discovered west of Dawson's digs in the Cypress Hills, although it would be another twenty years before anyone realized what it was he had found. He came upon them near Drumheller on June 9, while looking for (and finding plenty of) coal. He'd been given a huge square of prairie and badlands to survey by the Geological Survey, something like forty-three thousand square miles, and it was entirely prescient of him, when he found the huge skull stuck in the cliffside like a plum in a pudding, to put aside his sextant and spend a whole afternoon taking it out, using only an axe and a geology hammer, then to take a week carting it to Calgary on the back of a wagon. This feat always makes modern paleontologists shudder when they think about it. A lot of delicate material must have been smashed by axe blows or broken off as the unjacketed skull was lowered by rope down the incline. He went slowly in the wagon, but he had neither glue nor plaster nor burlap, and more bone must have disintegrated on the bed of the jouncing wagon. What was left was sent to Ottawa, where it languished, and Tyrrell went back to his surveying and lignite. He left the Survey shortly after, and went into private practice in the Yukon, where he died a wealthy man.

Five years after his spectacular find, another fossil hunter, Thomas Chesmer Weston, discovered a second skull, not far from the first, and sent it back to Ottawa, where it, too, languished. Edward Drinker Cope examined both specimens in Philadelphia in 1898 and declared them to belong to a carnivorous dinosaur he had named *Laelaps incrassatus*, a species that was later deemed not to exist, and so in 1904 the Geological Survey's Lawrence Lambe took a look at them. He put them in with something he called *Deinodon*, a genus established solely on tooth shape, but that classification was also given up when it became apparent that a lot of very different dinosaurs had very similar teeth. Finally, when more material started showing up in the United States, Henry Fairfield Osborn grouped it all together in 1905, the year Alberta became a province, and called it *Albertosaurus*, a close relative, he said, of *Tyrannosaurus rex*, which he also named that year.[2]

Albertosaurus was a large theropod, though not so large as *T. rex*. It grew to about twenty-five feet, and weighed up to three tons, easily big enough to cause serious damage to a migrating hadrosaur herd, which was its chief food source. And, as the Dry Island quarry

---

2 *T. rex* was actually a direct descendant of *Daspletosaurus*, a theropod so far found only in Dinosaur Provincial Park. At one time, *Daspletosaurus* and *Tyrannosaurus* were thought to be one species, and *Albertosaurus* and *Gorgosaurus* were similarly lumped together. Lately, Phil has separated them into separate genera. *Gorgosaurus* and *Daspletosaurus* are found in Dinosaur Park; *Albertosaurus* is found in Dry Island; and *T. rex* is found in the more recent Frenchman Formation.

suggested, it hunted in packs. Evidence of pack behaviour in theropods was mounting fast: a *T. rex* quarry in the U.S. was found to contain four adults; the *Giganotosaurus* bone bed we had worked in Patagonia had had half a dozen animals in it, and when Barnum Brown discovered this *Albertosaurus* site in 1910, on a fossil finding expedition for the American Museum, he found nine individual *Albertosauruses* in it, and may have speculated even then about pack behaviour. Phil was in no doubt that we were dealing with a pack; what he wanted to see was whether the dinosaurs in it were all adults, or if there were juveniles and possibly females in the group.[3] He also wanted to know how they died. It was the same question he had asked in Patagonia: What would it take to kill nine four-ton predators all at the same time? A flood of biblical proportions? A drought? "One possible cause is forest fire," he wrote in a book for children published in 1998, before re-opening the Dry Island quarry. It is difficult to determine cause of death when the clues are seventy million years old, however. But burned wood can be preserved as charcoal in the fossil beds, and that, wrote Phil, "is something we will be looking for as we excavate more of the *Albertosaurus* bones."

Barnum Brown found this quarry during one of his famous drifts down the Red Deer River on a twelve-by-thirty-foot raft big enough to hold a cook tent and an area for fossil preparation, with deck space left over for stacks of crated bones. Brown and his two assistants floated down the river from Big Valley to Willow Creek, scanning the hillsides as they passed, and when they came to a promising site they would pole over to the bank, tie up, and make a temporary camp while they explored the area for dinosaurs. He'd opened this quarry that first year, took out a few bones, then got

---

3 It seems that the big theropods were matriarchal, since the females were generally larger than the males. It's usually difficult to determine a fossil's gender, but in the case of *Albertosaurus* the females had one tail vertebra fewer than the males. The missing vertebra was the one at the base of the tail, which presumably was lost to make birthing easier.

sidetracked at a richer bonefield downriver and never came back. Over the next three summers, he collected hundreds of tons of specimens, floating them downstream usually to Steveville, where a man named Steve kept a tavern and operated a ferry, or to a dirt track leading to Patricia, a small village from which the crates could be transferred to a cart and taken to Calgary. From there they were shipped out of the country to the American Museum of Natural History.

That's where Phil found them. Off and on, he'd been rooting around in the AMNH's basement for several years, looking for clues to unmarked quarries, double-checking unidentified material, and mentally going over ground that hadn't been examined since the early 1900s. While looking through Barnum Brown's 1911 drawer, he found a collection of small bones marked "tyrannosaurid" that he thought had surely been misidentified; they looked more like ornithomimids, bird mimics, to him. "Then I compared them to bones in another drawer," he said, "and realized, 'No, these are actually baby tyrannosaurs.'" The fact that young *Albertosauruses* (members of the tyrannosaur family) so resembled birds that they could be mistaken for ornithomimids dramatically underscored the link between theropods and birds.

AFTER WE'D BEEN hiking for about half an hour, Phil pointed to a high butte still quite a distance ahead of us, separated from Dry Island by a deep chasm. About halfway up one side, the vertical face levelled off slightly, forming a small shelf. On it was something white, like a tent, flapping in the wind. As we got closer I saw that it was a sheet of clear plastic stretched over a moveable wooden frame, a makeshift shelter that could be dragged over the bonebed to protect the exposed specimens from the rain. It was slightly more elaborate than the tarp we'd stretched over the quarry in Patagonia, but it served the same purpose.

When we finally stood in the quarry it turned out to be a narrow shelf of rock no more than ten feet wide and forty feet long, twenty feet down from the top of the butte and several hundred feet up from the valley floor. We gazed east over the brown thread of the Red Deer River where it carved its meandering course through the Badlands, far below us. The tops of evergreen trees swayed distantly beneath our feet. The Cree had used the cliffs opposite as a buffalo jump, driving huge herds of bison over the edge and carving up the carcasses for meat at the bottom. An anthropologist once told me that at other, shallower buffalo jumps the talus slopes at the bottom were rich in Indian artefacts, obsidian blades, arrowheads, clay pots, because after the beaters at the top had driven the herds over the edge, hunters below had to finish the animals off, the fall not being enough to kill the buffalo outright. Not here at Dry Island, she'd said. Archeological sites here yielded very little weaponry, because no buffaloes survived the eight-hundred-foot drop. All the Cree had to do was cut up the carcasses, drag the meat over to the fires to smoke and dry, and roll it up in its own hide for winter storage. Now, looking across the valley at the cliff opposite, I pictured a herd of several hundred bison spilling over the edge, the dust and the thunder, a solid, brown cataract of flesh breaking up into individual shapes as the animals tumbled down the steep slope and splashed into the water below. The river must have swelled with carcasses and blood. I even wondered if that was how the Red Deer River got its name.[4]

I KNEW A GLIDER pilot who told me that air to him was like water to a fish. When he glided, he sailed on currents and eddies, through a multi-levelled sky, each layer having its own signature of

---

4 It wasn't. The Red Deer River was called the Elk River by the Cree and Stoney Indians, because of the many elk to be found in the Badlands. The early Scottish factors looked at the elk and saw red deer, animals native to Scotland but unknown in the New World.

temperature and density. When he stood on the ground and looked up, he said, he felt as though he were standing at the bottom of an ocean of air.

Looking out from the *Albertosaurus* quarry reminded me of that. We were about halfway between the bottom of the valley and the top of Dry Island, which was a body of land completely surrounded by air. When we filled a plastic bucket with rubble and tipped it over the side of the quarry, exactly as we had done in Patagonia, it was all we could do to keep from jumping after it. Stepping off the edge would have seemed as natural as going for a swim. Once, when I dumped a bucket, I stood pondering the river far below. It suddenly seemed close, as though I was standing on its banks, that if I reached out my foot I would step into it. Then I noticed a small, white form moving above it. Through my binoculars I saw that it was a pelican, a huge bird, almost the size of a swan. It flew lazily down the river, scanning the muddy surface for fish. From where I was it looked like a tern or a small gull.

Once, in the south of France, near the city of Nîmes, my wife and I visited the Pont-du-Gard, an ancient, stone aquaduct that in Roman times carried water down the Rhône valley to Marseilles. That summer we just stepped out onto it and walked across. Merilyn stayed inside the aqueduct, where the water would have flowed, while I picked my way along its flat top, from where I could see far up and down the wide river valley. About halfway across, directly over the deepest part of the river, I stopped to look down. I was standing on a limestone slab cut and fitted almost seamlessly to other slabs by Roman workmen two thousand years before. Embedded within the rock were fossil shells I knew to be millions of years older. And as I knelt to examine them, I saw a set of initials, "J.M.M.," carved above a Masonic symbol, compass and set square, into the stone among the shells, and a date: 1735. My head swam with a kind of dizziness that was unconnected with the bridge's height above the river. It was temporal vertigo, a feeling that I was falling not through space but through layers of time, swirling through the present, through the eighteenth century, through the Roman era, through the unchartable epoch of fossil shells, to a time even older than that, when rock is scraped and crushed into mud, then squeezed and heated and concentrated back into rock. Suddenly the sound of a bell returned me to the actual present. Peering over the edge of the aqueduct I saw three greyhounds loping along the riverbank. One of the dogs had a goat's bell tied to its collar; I could hear its rhythmic clanging echo mournfully in the walled valley. I felt as though I were running with the greyhounds through a sort of perpetual present. Now that solitary, languidly flying pelican, swimming through the air below me, floating on an invisible river of time, brought that eerie orientation upon me again. Time collapsed, condensed like rock until time and bone and rock were one. I thought: When the pelican took off from the water, somewhere upriver beyond my field of vision, it had been a dinosaur. And when it continued, following the river's flow around the next bend, it would change into some other form of being, one that also already exists.

MOST OF THE VOLUNTEERS were working under Darren's supervision in the soft side of the quarry, where a large number of disarticulated theropod bones had already been exposed by erosion. Barnum Brown had brought the quarry down to bone level, working fast and carelessly, breaking bones he didn't want to get at the ones he did. He counted nine *Albertosauruses* but collected only the hind feet, some jaws and legs, and dumped the rest in a midden or spoil pile at the south end of the bed. When Phil reopened the quarry in 1998 with a volunteer group similar to ours, they upped the *Albertosaurus* count to at least twelve. There was also a hadrosaur in there somewhere, a *Hypacrosaurus*: we kept coming up with its teeth, curiously squared-off, tubular pegs mixed in with the theropod material. And next to the spoil pile we found something that looked at first like a fossil tree stump, a foot in diameter with branches radiating from it, but which turned out to be five articulated hadrosaur vertebrae, complete with five flat spines, curving up from the centrums like a rack of bread knives, each about two feet long. The whole assembly, five vertebrae and their five spines, was known as the sacrum, the heaviest part of the backbone where the pelvis attached, the animal's centre of gravity. Normally Phil would not be excited about a partial hadrosaur, but a complete sacrum was rare, and a *Hypacrosaurus* mixed in with a dozen *Albertosauruses* was more than just another hadrosaur; it was a story.

So he wanted to take out the sacrum. The question was: how would we get such a huge jacket out of the quarry, back along the slippery trail and up the final cliff to prairie level? Charles Sternberg, who quarried the Badlands in the 1920s, took out a hadrosaur sacrum similar to ours; it weighed more than six hundred pounds, and required pulleys, a horse and wagon, and weeks of road work to get out. Phil thought about getting it down to the river somehow, loading it on a barge, as Barnum Brown did with his specimens, and floating it out that way. But getting this one down would be hardly less dangerous than getting it up. Then the British Army came to our rescue; the British had a helicopter training base nearby, and offered to lend Phil the use of a chopper and pilot to fly the heavier jackets out. It would be good training for the pilot, they said. So Phil's plan was to jacket the sacrum, sling it in a rope net, and let the helicopter make it fly.

I was assigned to the Hardrock Café with Stewart and Jørn. Here the matrix was mostly fine-grained and extremely hard, bluish grey in colour, fossilized potter's clay. Embedded in it were dozens of dark brown bones, like soupbones floating in a cloudy aspic. Stewart handed us each a pair of ear plugs and goggles, and bent to start up a small, gas-powered compressor with a network of red rubber hoses running from it. At the end of each hose was an air scribe, a metal handle equipped with a sharp, vibrating tip that disintegrated rock by turning it into powder, one grain at a time. We set to work. The scribes were easier to use than hammers and chisels, and safer for the bones, but the noise was deafening, even with ear plugs, and escaping air blew stone dust into our nostrils and throats.

I had a rib to take out, not one of the longer hadrosaur ribs but one that seemed to be about a foot and a half long. The pointed, or distal, end was exposed; the proximal end that would, in life, have attached to the vertebra, disappeared into a shelf of hardrock. I lay on my stomach and went to work. First I removed the rock above and behind the rib, so that the entire length was exposed at the bottom of a basin carved into the rock shelf. I then removed more rock from behind it until it lay in a cavity about the size and shape of half a bushel basket. Then I trenched around the rib and undercut it so that, eventually, it stood about six inches up from the bottom on a narrow pedestal of rock, ready to be wrapped in burlap and plaster before the pedestal was snapped off. Using the air scribe, I had turned about half a bushel of rock into powder, a process that took me the whole of the first day and part of the second.

THE NEXT MORNING, as I was completing the rib, I heard a shout from the top of the cliff, directly above the quarry, and when I looked up, a man with a black hat, white shirt and red suspenders was looking down at us, outlined against the grey sky like a tall portent.

Stewart stood up and shut off the compressor. The man above us was joined by a woman, then two small boys and a girl of about thirteen. The woman and girl were wearing long, blue calico dresses with grey aprons and bonnets. The boys wore black pants and red flannel shirts.

"Holy cow," Darren said to Phil. "It's the Hutterite man I told you about, the one who said there were no such things as dinosaurs."

Phil raised his eyebrows and grinned. "Cool," he said. "Invite them down."

When the family scrambled down from the top of the cliff, the man repeated his question: "Have you got it out yet?"

"Not yet," Darren said. "We've had to move a lot of dirt first."

The man smiled kind of twistedly, and he and his family walked around the soft side of the quarry without stepping into the bonebed. The boys were twins. They squatted at the edge of the quarry a few feet from the exposed bones, and with their hands tucked under their bellies and their elbows sticking out they looked like two black and red coot chicks. They watched closely as Andy and Brooke brushed dirt off an *Albertosaurus* jaw, exposing a row of five beautifully preserved, marbled teeth that protruded from a curvature of bone. Beside it lay gigantic leg bones, a tangle of metre-long ribs, part of a pelvic girdle.

"How do you know they aren't buffalo bones you're digging up?" the man asked. "Or maybe just some old cow bones." His expression, as he looked at Darren, was serious. He was not saying: "These are just cow bones." He was asking: "How do you know?"

Darren told him that, first of all, they were not bones any more, that the process of bones turning into rock took a long time, longer than there had been cows, longer even than there had been buffalo. "Millions of years," he said.

"How many millions?"

"For these," said Darren, "we think about sixty-eight million years."

The man shook his head, not in disbelief but as though to clear it. "What do they look like when they're all put together?"

The quarry book, which contained a drawing of every bone known to have come from an *Albertosaurus* and a sketch of a complete skeleton, hung from a nail in the plastic tent. Darren took it down and showed it to the man. "It was a big dinosaur," he said. "It was maybe twenty-five feet long and weighed three or four tons. You could see what it looked like if you came down to the Tyrrell Museum in Drumheller."

The man nodded. "I've been to your museum," he said. "It's easy to fake a dinosaur. You could just carve it out of rock."

"No, we couldn't," said Darren. "Rock looks different inside. Besides, why would we do that?"

The man was quiet. Why would so many people go to so much trouble just to prove that the Creationist universe was wrong? Well, people have gone to a lot more trouble for a lot less. But he didn't say so. The Bible speaks of giants on Earth, but they were clearly men, not beasts, and God sent the Flood to destroy them. He didn't say that, either. The woman and girl looked around the quarry at us, and we kept working, letting Darren field the questions. Phil stepped carefully, holding his notebook, adding new bones to the quarry map while Eva sat cross-legged beside a small mound that contained a cranial bone. The boys took their eyes from the quarry only to look at their father, whose face maintained a lively but noncommittal interest.

"How do you know what kind of dinosaur they came from?" he asked.

"If you found a truck door by the side of the road," Darren said, "would you know if it was from a Ford or a GM?"

"Yeah, sure," the man said, laughing. "The Ford door would be all rusted."

"Well, we've studied dinosaurs as much as you've studied trucks. Here." Darren turned a page in the quarry book. "Here is a dentary, a lower jaw," he said, and then pointed down to the quarry floor. "That's this one." The *Albertosaurus* dentary lay brown and gleaming, like a chestnut, against the sandstone floor. "You can see we haven't faked that, can't you?"

"Yes, that's a real bone," said the man.

"And here, here is a tooth." Darren handed him a tooth that Phil had taken out that morning. It lay like a polished onyx railway spike in his hand. The boys stood on tiptoe and peered into their father's hand, as though they would drink. "I think you'll agree that that's no cow tooth. You can see the serrated edges. That's the tooth of a carnivore."

The man held the tooth for a long time, nodding his head silently, turning it over, turning it also in his mind. Then he gave it back to Darren, took his sons' hands, one in each of his, and moved towards the path, his wife and daughter in the lead. At the top of the cliff he stopped and turned around. He seemed to have come to a conclusion.

"I didn't believe in dinosaurs when I came here," he said. "But I do now." The boys stopped and looked up at him. "Thank you," he said.

NORMALLY AT DINNER Martha would put everything out on the table in the kitchen trailer, and we would file in with our plates and help ourselves. But on July 1, Canada Day, Martha took a break, John H. cooked steaks on the gas barbecue outside the trailer, and John A., the lugubrious waiter from Yellowknife, dressed up in a black tuxedo, complete with ruffled shirt, cummerbund, white gloves, blue jeans and running shoes, and waited on us at the picnic tables.

"And how would you like your steak, sir?"

He poured the wine, sampling each glass first: "Ah, an excellent month," he said, then added, "A vintage joke, and a fine one, don't you think?" His walk became steadily more unsteady as he balanced plates on his fingertips from the barbecue to the tables, so of course we called him Sir John A. The vegetarians among us he fixed with a solemn, long-suffering regard. "Well, then, how would you like your potato?"

For thirty years, he'd told me earlier, his father had owned the Canadian Tire store in Yellowknife. John A. had worked in the store with his father for a while, and had been expected to take it over when the old man retired, but he'd ended up waiting on tables for six months of the year and drafting mining maps on a computer for the other six. He lived, he said, in a reconditioned houseboat moored on Great Slave Lake, and when he didn't feel like waiting on tables or drawing maps any more he called in sick and stayed home to reorganize his fossil collection. "You can buy almost anything on the Internet these days," he told me.

There is something about the north that causes people to shed their work ethic, like dead skin, when they cross the sixtieth parallel. The Yukon and Northwest Territories are our Patagonia; perhaps it has something to do with distance from the capital. Gilberto de Mello Freyre, a Brazilian social philosopher, maintained that the increased leisure made possible by advances in technology was making North Americans more like South Americans in their attitude towards what was important and what was frivolous. We, that is North Americans, were becoming more social and therefore more cultural, more like the Latin, that is the Iberian, south. "Extreme idealization of toil has become an archaic tendency," he wrote in 1963. "It is not the Iberian conception of time that is now archaic [to North Americans], but the Anglo-Saxon conception that went to the extreme of identifying, not only in Europe but in the imperial activities of Europeans in Africa, Asia, and America, time with money."

Freyre seems to have worried that all this leisure time we were about to enjoy would be wasted on us non-Iberian types. In an essay entitled "Toward a New Leisure," he noticed that we were exhibiting "a growing, nonsectarian preoccupation with the future: a future which is already leaving its mark upon the present." He thought we should channel this thinking about the future into something useful, something that would fill our thinking "with rich material of new significance for man's existence and new and mean-ingful motivations for human action." But wasn't that exactly what all these technological advances were supposed to free us from? Ever since the Industrial Revolution, Freyre opined, our creative energy has been subverted into labour by "first, the lords of com-merce and, more recently . . . the kings of industry." Now that we have all this leisure time in which to think about the future, "the lords are dead, and the kings may already be dying."

There didn't seem to be a surplus of leisure in our camp, but then we were thinking more about the past and the present. But Freyre might have a point, since John A., sitting leisurely in his tethered houseboat, surrounded by his fossil turtle shells and thero-pod phalanges, was thinking about the future.

"Is there a Canadian Tire in Drumheller?" he asked me one day as we were working in the quarry. And when I told him there was, he said: "I think I'll drop in to see the manager the next time we go into town."

IT WAS BECOMING clear what had happened. The quarry con-tained at least a dozen *Albertosauruses* and one *Hypacrosaurus*. There did not appear to have been a forest fire. Sixty-eight million years ago, when this quarry was a river estuary on the shore of the great Western Inland Sea, a pack of carnivorous *Albertosauruses*, mostly adults, a few adolescents, perhaps still in their first downy plumage, and one possibly terrified *Hypacrosaurus*, were crossing a river some distance inland, in what are now the foothills of Alberta. The *Albertosauruses* might have been chasing the *Hypacrosaurus*, or perhaps the *Hypacrosaurus* was crossing the river at a different point, we don't know. But halfway across they were all caught in a flash flood, a huge wall of water charging down the channel from something happening farther inland. Since this is speculative nonfiction, or scientific fiction, let's make it a rainstorm, the kind of torrential downpour you get when mountains rise up through time and poke holes in the liquid air. The water caught the animals by surprise and swept them downstream. *Albertosauruses* were not a species accustomed to fear, but the *Hypacrosaurus* certainly was. There may have been a lot of snarling mixed with braying or honking; there may have been, for the first time, terror in the *Albertosauruses'* eyes.

It was a big pack, let's make it a dozen. The biggest animals, the females, might have been teaching the juveniles to hunt. There might have been an atmosphere of play in the pack, a lowering of the guard. I see them charging into the muddy water, perhaps in pursuit of prey. They were reptiles, they normally might have liked water, but this was different. They might have snapped at passing logs with their teeth as they drowned, they might have snapped at the *Hypacrosaurus* and at each other, but nothing helped. By the time they were carried downriver to the coast they were all dead, predator and prey, the water, gentler now, thick with silt, laying them all down together on a sandbar and lapping them with mud, the lamb nestled with the lions, the questers with their quarry. Slowly their flesh rotted and their bones were buried; gradually, so gradually that anyone watching the process would have to be unaffected by the passage of time, they turned to rock.

CHAPTER TWELVE

The Fossil Song

A DAY OR SO LATER, while the rest of the crew began jacketing the specimens for the helicopter lift, I left Dry Island to go to Dinosaur Provincial Park. As I drove south I was again travelling back in time, for the formations in Dinosaur Park are about five million years older than those in Dry Island, and contain an older dinosaur assembly, in some cases ancestral to it: *Gorgosaurus* and *Daspletosaurus* instead of *Albertosaurus*, *Lambeosaurus* instead of *Hypacrosaurus*.

On the way, I passed through Drumheller again and this time dropped in to the Tyrrell to see Don Brinkman, a fossil turtle specialist whose two main goals in life were to figure out how, when so much reptilian life on Earth was lost at the end of the Cretaceous Period, turtles managed to squeak through unscathed, and why more paleontologists weren't interested in that question. I'd got to know him in China when, in the Gobi Desert, I'd found a skull from an unknown species of Cretaceous turtle. I didn't know it was a turtle skull I'd found, of course. I thought it might have been a pelvic bone from some larger reptile. It was about the size and colour of a softball, with two holes on one side and a kind of flaring at the base, like the back of Darth Vader's helmet. When I took it out, cleaned it up with my Swiss Army knife and showed it to Phil that evening in camp, his jaw dropped. "Show this to Don," was all he said. That night at dinner I put the skull on Don's plate, the matrix-filled eye sockets peering broodily at his empty chair, and when he came in and realized what he was looking at his face lit up with surprise and genuine pleasure. He picked it up and turned it in his hand as though it were a precious stone, which in a way it was, and passed it around the table.

Don's office in the Tyrrell was filled with so many fossil turtle shells it looked like the studio of a potter obsessed with the perfect bowl. He was born in Alberta, near Drumheller, went to McGill to study paleontology, then spent three years studying (fossil) pelvic structures in Harvard's Museum of Comparative Zoology before coming to work with Phil at the Tyrrell. When he got here, "so many people were working on dinosaurs there didn't seem to be room for one more, so I looked around to see what wasn't being studied and came up with turtles. So I made turtles my specialty."

Don knew that turtles weren't as sexy as dinosaurs. No one really cared about turtles. There was never going to be an international symposium on turtle evolution, as there had been on dinosaur extinction, or warm-bloodedness, or the origin of birds. Turtles hadn't changed very much. There were no science-fiction movies in which humans are chomped by demented, rampaging turtles. Turtles were dull and plodding and not particularly intelligent. They hadn't been wiped out by giant meteorites or sudden climate changes. They didn't come in a variety of shapes and colours. They didn't have feathers. But Don liked them anyway. "I think they have a lot to tell us about life on this planet, if only by the very fact that not much has happened to them over geological time. They haven't changed much because they haven't needed to change much. They got it right the first time. Maybe we should be looking at that."

THE FEATHERED DINOSAURS from China were at the Tyrrell. I had seen them at Yale's Peabody Museum in February, during the

Ostrom Conference, when along with two hundred paleontologists, including Larry Martin and John Ostrom, I had lined up and filed past their glass-fronted cases. I wondered then what those two men had thought; both looking at the same markings, Ostrom seeing feathers and Martin seeing frayed subcutaneous scales, Ostrom basking in the apotheosis of nearly thirty years of research, Martin sensing the pendulum pause and begin to swing back towards himself. And now I stood in line to see the specimens again, this time with a group of school kids and their parents. The tiny dinosaurs occupied a small room of their own, just inside the turnstiles, past a television set showing endless repetitions of Discovery Channel's *If Dinosaurs Could Fly*.

They were presented on six stone panels, mounted behind glass in six wooden cases. One contained two perfect specimens of *Confuciusornis*, the only one of the feathered dinosaurs that could fly; it had claws on its thumbs and middle fingers, but its index finger was composed of long, flat bones to which the flight feathers attached, something that had not been apparent on the unfinished specimen Phil had shown me last January in the prep room. Although it has been called both a dinosaur and a bird, *Confuciusornis* does not quite qualify for missing-link status, since its wing feathers were longer than its body, indicating that, like *Archaeopteryx*, it was following an evolutionary flight path that did not lead to modern birds. But its wing feathers definitely looked like feathers, not frayed scales.

Two other panels contained a juvenile and an adult *Sinosauropteryx*, the first of the true feathered dinosaurs found in 1996. The juvenile was without flight feathers, but its body, all along the spine and up the long, straight tail, was covered with a short, downy frill, which suggests to evolutionists that feathers first proved useful for warmth and only later for flight. The adult, though more fully feathered, was flightless, clearly a dinosaur, having pointed, theropod teeth with serrated edges. Its lower jaw was not fused together at the chin, as it is in modern birds: *Sinosauropteryx* could clamp its mouth over a struggling prey and have enough give in its head to prevent its teeth from snapping off. Lying where the theropod's stomach would have been were two tiny mammal jawbones, the 120-million-year-old remains of its last meal.

Caudipteryx, on the next panel, was curled in a typical dinosaur death pose, long neck arched, head thrown back, tail raised almost to meet it, in its gut a tiny sac of stomach stones, still preserved in the chalky matrix, that helped break down plant material. Like hoatzins, it was primarily vegetarian, although it might also have eaten insects. A second specimen showed short feathers spread out and attached to the manus, like a hand of cards, too short for flight. What were they for, then? What purpose would long arm feathers serve before they would prove useful when it came time to fly? A second fan flared from the end of the long tail, also not for flight. Sexual display? Did male *Caudipteryxes* and *Sinosauropteryxes* strut about like tiny flamenco dancers, their faces covered coquettishly by their wrist fans and, in the manner of peacocks or sharp-tailed grouse, spread their colourful tail feathers to attract the notice of females?

Finally there was the supremely enigmatic *Protarchaeopteryx*, the only specimen so far, with its huge legs, short arms and tiny body. It looked like a powerful runner, not unlike an ostrich; Ji Qiang had named it *Protarchaeopteryx robusta*. It had serrated teeth that resembled those of *Archaeopteryx*, a clump of feathers at the end of its long tail, more along its spine and down its legs, and a furcula. It couldn't fly either: it was, in fact, less evolved—more dinosaur-like—than *Archaeopteryx* even though it lived twenty million years more recently.

No one who really looked at these specimens would ever see birds in quite the same way again. Not only have they augmented our understanding of dinosaurs, but they have also changed our definition of "bird." We moved silently through the small room as though we were in a cathedral. The physical walls around us had been expanded by the presence of these delicate creatures. In New Haven, the atmosphere had been charged with mute awe. Here I had been expecting noise, yanked arms—C'mon, Daddy, let's go see the dinosaurs. But there was none of that. Wendy Svoboda, one of the Tyrrell's preparators, sitting in a corner behind a glass partition, was demonstrating how these specimens had been prepared, but no one seemed to notice she was there. The overpowering beauty and magnificence of these tiny fossils completely filled us. They were not huge, menacing, unimaginably evil beasts with no direct connection to anything we knew, fantasy creatures, nightmare monsters; we could not look upon their bones smugly, as we do upon *Albertosaurus* and *Tyrannosaurus rex*. When we contemplate the fact that those other, ferocious carnivores are safely extinct, we do so with relief and a certain superiority: See, we say to ourselves, how the mighty have fallen? But these little animals, these pigeons and bantam roosters, these pygmy owls, hushed us in an entirely novel way. See, they said to us, this is how the humble have fallen. This is the way the world ends.

THE FACILITIES AT THE Field Experience camp in Dinosaur Park were even posher than those at Dry Island. Here the kitchen trailer was big enough to hold not only the kitchen but also four rows of tables and chairs, a sort of dining room in which, as I entered, half a dozen new volunteers were scanning a large blackboard for their assignments. Outside, arranged around a square of recently sodded grass, were five more trailers, four containing sleeping quarters for volunteers and staff, and one, across from the kitchen, with two shower rooms and a laundry. Three people were playing competition-calibre Frisbee on the grass. There was a between-sessions atmosphere about the place that reminded me of the times I'd dropped my daughters off at various summer camps, with the previous groups having finished their tearful farewells and the new fish trying to appear relaxed and at home. Here, though, instead of a reedy lake and a few battered aluminum canoes pulled up on a muddy beach, there were the almost limitless Badlands and four grey Jeep Eagles parked on a gravel lot; I could almost sense their impatience, as though they were teams of horses harnessed and stamping in their eagerness to go. Beyond the trailer complex, on the banks of the Red Deer River, was the VIP campground, a large meadow with a circular road through it giving access to a ring of campsites. Mike and Don had told me that I could park the van there and live in it if trailer-park life didn't appeal to me, and that's what I decided to do. I had the campground to myself: I must have been the only VIP in camp. I parked in the site closest to the river, so that my view through the sliding side door was upstream, into the morning sun, and looking deep into the Badlands along a sluggish curve of chocolate-coloured water whose low banks were lined with overhanging cottonwoods and the silver sheen of wolf willow. I turned off the engine, got out and sat at the weathered picnic table. Western kingbirds, perched on tall stocks of dried weeds, leapt suddenly into the air and hovered like helicopters amid swarms of swirling insects, snapping at them with their reptilian beaks. The river flowed by so thickly it seemed to have been poured, like chocolate cake batter, from somewhere higher in the hills; above it, gulls and bank swallows flitted and swooped, and from across it came the raucous squawk of a heron. I wondered if that ancient, primaeval cry had also echoed in this canyon air some seventy-two million years ago from a similarly reptilian throat.

Back home, I rarely noticed grass, but in the West, grass invades the senses in such a variety of forms that it is almost inaccurate to refer to it in the singular: the word is grasses. Any given meadow, even one like this campground at the bottom of the Badlands, is a meeting place for several species, each with its own poetically descriptive name: awned wheatgrass, green needlegrass, prairie dropseed and tufted hairgrass. The dominant form in my campsite seemed to be a tall, panicled, rhizomed and long-bladed *stipa* called speargrass, also known as needlegrass, or needle-and-thread, or porcupinegrass, whose barbed, quill-like heads shot through my socks and clung to my ankles for dear life. There were also prickly pear and pincushion cacti, three kinds of wild sages—silver, pasture and prairie—and a variety of broadleafed plants, including the lovely scarlet mallow and the less lovely skeleton weed.

At a meeting in the kitchen trailer that evening the volunteers and leaders gathered for a brief briefing. We took beers from the fridge, a new Western brand called Magpie, or coffee from a huge urn, and sat at the tables in comfortable expectation. Brooke, Andy and Al had also come down from Dry Island, and Don Brinkman and Mike Getty I knew. We were divided into three groups: one was to work in a ceratops quarry not far from camp; a large theropod, probably a *Gorgosaurus*, was ready to be taken out in a more distant part of the park; and a third group, to which I was assigned, was to go prospecting with Don. Prospecting was just walking around in the Badlands looking for bones, and after two months of stationary quarry work in Patagonia and Dry Island, I liked the idea.

After the meeting we went to the Tyrrell Museum Field Station, a sort of mini-museum near the entrance to the park containing a small display area and gift shop, a large preparation room and a suite of park administration offices, including a library. In the prep room, a vast, high-ceilinged chamber with shelves of plaster-wrapped dinosaur parts waiting to be cleaned and readied for analysis, Mike handed out field kits containing a geology hammer, a set of awls and dental picks, a paint brush, and a plastic bottle of

Vynac, a glue-like preservative that prevents air and water from getting at newly exposed bones. During the day, from eight to five, we would work in the park, in the quarries or on the prospecting teams, he told us, and in the evenings we would come back to work in the field station on previously collected specimens.

Later that evening I went into the library and read a *National Geographic* article by University of Chicago paleontologist Paul Sereno, about his collecting adventures in Africa. I'd seen Sereno at the Ostrom Conference at Yale earlier that year. He was a well-groomed, vigorous man who had spoken about the shared characteristics between birds and the family of small theropods known as the Coelusauridae, and about where *Caudipteryx* might fit in that lineage. He had been a research associate at the Royal Ontario Museum in the early '90s, and had spent four field seasons in Argentina before becoming interested in Africa. In 1993, he'd taken a team to Niger, a desert republic sandwiched between Algeria and Nigeria, and been caught up in that country's "political turmoil," as he called it in the article. Just as Darwin and George Gaylord Simpson had found themselves surrounded by revolution in Argentina, and as Phil and the other members of the Dinosaur Project had been caught in northern China during the Tiananmen Square massacre, so Sereno and his crew drove into civil war in Niger. The bonebed he wanted to visit was in a military zone, and he had neglected to get the proper papers. He and his crew languished in Agadez for seven weeks, under house arrest, before they got permission to proceed. With only half the field season left, they travelled to a place called In-Gall, where they found a fifty-five-foot sauropod and a thirty-foot theropod Sereno named *Afrovenator abakensis*, which means the African hunter from the In-Abaka oasis. The interesting thing about both specimens, apart from the drama surrounding their discovery, was that they were 130 million years old, and yet seemed to be related to the North American dinosaurs *Camarasaurus* and

Allosaurus, which disappeared 150 million years ago. In order to know why that was interesting, it helps to know a little about continental drift.

During the Triassic Period, the first of the three dinosaur epochs, the Earth's land mass was clumped together into one supercontinent called Pangea. Just before the Jurassic, the second epoch, the whole thing started to split in half across the Equator, like a macrocosmic illustration of cell division. Half the land drifted north and was called Laurasia, the other half went south and was called Gondwana. Until the end of the period, Laurasia and Gondwana remained almost entirely unconnected: for the most part, dinosaurs in one evolved quite separately from those in the other. There was, however, a small land bridge between what today are Spain and northern Africa, which seemed to have allowed some traffic between the two land masses. Imagine the Strait of Gibraltar closing, and the Mediterranean opening up to become part of the Indian Ocean. Some dinosaurs may have crossed this bridge and intermingled with those on the other side.

In the Cretaceous, the third and last dinosaur period, Laurasia and Gondwana began to break up, and pieces floated around and bumped into one another. Laurasia became North America, Europe and Asia; Gondwana broke up into South America, Africa, Australia and Antarctica. The land bridge between Africa and Spain—now the Strait of Gibraltar—was thought to have disappeared 170 million years ago: Sereno's new dinosaurs suggested that it must have either remained in place until more recently or else have risen again some time between 150 and 130 million years ago. If Sereno could ascertain when African theropods resembled North American theropods more than they resembled, say, Argentine theropods, he might be able to reconstruct the time frame for these Cretaceous connections, as well as map dinosaur dispersal and migration from one landmass to the other. Very little was known about African dinosaurs: as far as paleontology was concerned, Africa was still the Dark Continent, and Sereno was a sort of modern-day David Livingstone.

Sereno returned to Africa in 1995, this time presumably making a few advance inquiries into nuisance items like travel permits and research clearances. He went farther north, prospecting in some Late Cretaceous deposits in the Atlas Mountains in the Moroccan Sahara, closer to where the Laurasia-Gondwana connection had been. There he unearthed "perhaps the largest carnivore that ever walked the earth," a huge theropod with a skull that was more than five feet long and a skeleton that measured forty feet from tail tip to snout. He identified it as a *Carcharodontosaurus*, the lizard with teeth like a shark's, a species that was first recognized from a few random bones found in Egypt in 1917. Sereno's specimen added enough information to allow him to guess at its size and family lineage: as near as he could tell, *Carcharodontosaurus's* nearest relative was the Early Cretaceous, North American theropod, *Acrocanthosaurus*.[1] But then Rodolfo found *Giganotosaurus*, which so closely resembled *Carcharodontosaurus* that Rodolfo placed it in the same family. If the *Acrocanthosaurus* connection held, there was now a continuous line of theropod relationships that ran from North America, through Europe into Africa, and across to South America, and stretching in time from Early to Late Cretaceous. "The dinosaurian world," Sereno wrote, "may not have split neatly in two."

Several other volunteers were sitting around the library table, talking quietly about their previous work with dinosaurs. Some

---

1 This is the large theropod whose footprints track those of a huge sauropod in the mudstone bed of the Paluxy River, in Dinosaur Valley State Park, Texas, eloquently telling the story of predator and prey. *Acrocanthosaurus* was a thirty-three-foot, 1.5 tonne dinosaur with tall, bony spines along its back, which made it look not unlike the *Dimetrodon* that first captured Phil's eight-year-old imagination. "In warm environments," notes Dale Russell in his book *An Odyssey in Time*, "a heat-radiating web may have been useful in allowing an animal to dissipate heat after running, thereby avoiding death by heat prostration." Dale himself narrowly avoided a similar death in the Gobi Desert in 1988, not by growing bony, skin-covered spines, but by being force-fed a gallon of salt water and rushed to a nearby hospital.

were on their first field trip, others were paleontology students looking to get some field experience on their CVs in preparation for post-graduate work and already had an extensive list of accomplishments. One of these was a young man named Rud, one of the Frisbee players I'd seen earlier. He was tall and evenly tanned, with chiselled facial features and long, fine blond hair that fell easily over his forehead when he talked. He looked as though he'd come straight from some California beach party, but when I spoke to him at the briefing session he told me he was from Chicago. Now, with a nod at the *National Geographic* I was holding, he mentioned that he had been to Africa as part of Paul Sereno's field team in 1997.

"We went back to Niger," he said, shuddering at the memory so that his hair slid across his eyes. Raking it back, he launched into the story of Sereno's 1997 field season. "I met Paul in New York," he began, "at that year's Society of Vertebrate Paleontology meeting. We hit it off right away, and sat up until two in the morning eating hotdogs and talking about dinosaurs. I'd been working as a preparator at the Field Museum in Chicago, on Sue, the huge *T. rex*. We'd work on her until midnight, then crash in sleeping bags in the Field Museum's basement collections room, then get up at dawn and start working on her again. All summer it was like that, exhausting, exhilarating, nerve-wracking. Paul told me about his 1993 trip to Niger, which had also been exhausting and nerve-wracking but also pretty exciting, and then asked me if I wanted to go back with him that summer, to Africa, and of course I said yes. Who wouldn't? Well, the trip turned into a four-month nightmare pretty quickly."

They landed in Ghana in September with four ship's containers of equipment, and after renting five trucks and three trailers headed in a convoy north into the Sahara, to Niger. "We had papers from the American embassy that Paul had had to sign. They were waivers, saying that we acknowledged we were proceeding into dangerous territory in disregard of the embassy's warnings, and that if we got into trouble the Marines would not be sent in to bail us out." They had hired armed guards to accompany them, fifteen Tuareg warriors who turned out to have been former rebels against the Niger government. "Trained killers," said Rud. "I'd never been so terrified in my life. I remember riding in one of the vehicles, on a really rough road full of potholes, with one of the guards sitting in the seat directly behind me. His rifle was resting butt-down on the floor, he had his finger on the trigger, and the muzzle was pointing through the back of my seat right at my head: I thought, one good pothole and I'm toast."

Despite his misgivings, Rud took to hanging around with the guards because he felt safer among them. "I even got them to show me how to operate a machine-gun," he said, and he showed us a photograph of himself standing against a sand dune with three Tuareg guards, all four of them carrying weapons and wearing belts of ammunition, Rud deeply tanned, his hair and neck covered by a Tuareg turban and veil, looking like a young T. E. Lawrence, with the same disturbing expression on his face, a curious and troubling blend of pride and unease. At one checkpoint, Sereno was dragged from his Jeep and held against a wall at gunpoint while his papers were examined; later, a tracer rocket was fired at them as a warning, but whether by rebels or government troops or just bandits they couldn't tell. Sereno pushed serenely on, Rud said: "He just worked on energy, he didn't stop to think. Several people bailed out, went home early. I considered it myself. One day I took him aside and said, Look, Paul, I'm worried, I'm really terrified, but I'm putting myself in your hands. If you say it's safe to go on, I'll stay. And he looked at me sort of startled, you know, as though he'd never thought of it that way before."

Sereno and his remaining team, including Rud, found another *Carcharodontosaurus* and a sauropod bonebed that contained several camarasaurid skeletons, including those of juveniles, which suggested that even sauropods may have travelled in herds, or at least that African sauropods consorted in family groups, somewhat like elephants. So much misadventure and mishap, I thought, to put together the rudiments of a theory of dinosaur behaviour. Hollywood poured millions of dollars into movies in which intrepid scientists underwent excruciating hardships to discover the whereabouts of some long-lost treasure, a fabulous diamond, a hoard of Egyptian gold. Indiana Jones was modelled on Roy Chapman Andrews, a real scientist who was sent to the wildest regions of Mongolia in search of the truth about human origins. Surely that was motivation enough, without having to throw in gemstones and golden scarabs to give his quest more credibility? If museums wanted to make their exhibits more entertaining, perhaps instead of building gigantic robotic *Iguanodons*, they could simply present the stories behind some of the fossils in their collections. As Rud could attest, the steady advance and retreat of human knowledge can be a white knuckler at times. When I asked him to name the most remarkable thing about the Niger trip, he replied instantly: "The fact that no one was killed."

THE OTHER NICE THING about turtles as opposed to dinosaurs is that they are small. Taking out a turtle shell, carapace and plastron, is a one- or two-person operation for an afternoon, not an expedition force of twelve that requires two months and a contract with the *Los Angeles Times*. Don Brinkman took three of us on a prospecting tour; me, Sarah, a schoolteacher from Seattle who, with no sign of self-consciousness, was wearing a straw lampshade on her head as though it were a garden hat, and Tammy, a home-care worker from Edmonton taking a break, she said, from lifting eighty-year-olds out of bathtubs. We drove to a remote section of the Badlands, and the first thing we found was a turtle shell. This was not particularly surprising or suspicious to me, since it is a feature of prospecting that you find what you're looking for. In China, when I prospected with Phil we found small theropods; with Dale, who at the time was interested in sauropods, sauropod bones littered the landscape; with Dong Zhiming, who was working on a paper suggesting that stegasaurs originated in Asia and migrated to North America, we saw stegasaurs everywhere. And when Don was around it rained turtles. The corollary of this axiom is that it is often frustrating to go prospecting without some idea of what you are prospecting for. You can't just wander around looking for anything interesting that comes along. A good prospector seems to coax specimens out of the rock, like a bone detector, and to do that you need to have a clear idea of what you want to see before you see it. That explains why someone like Phil can prospect the same area again and again, and come up with important finds with alarming frequency; each time he searches an area, he adjusts his mental scanners and looks for a different type of specimen.

We parked the Jeep at the edge of the prairie, climbed between the strands of a barbed-wire fence meant to keep cattle from making impromptu buffalo jumps, and filed down the ridge of a coulee into the Badlands. At first the hiking was easy; it hadn't rained for a while, and the clay was hard underfoot. Very soon, though, we were surrounded by looming, soft-sided cliffs of grey bentonite— ancient volcanic ash turned to clay, slumped and wrinkled into forms that always reminded me of sleeping elephants with their legs tucked under their enormous bellies and their trunks stretched out towards the river.

In Robert Kroetsch's novel *Badlands*, William Dawe, leader of the fictitious Dawe Expedition, scans the cliffs as his flat-topped raft drifts downriver past Steveville into "the deepest and widest section of the Alberta Badlands," that is into Deadlodge Canyon. When he stops to prospect he tries to explain to his crew members how to recognize bone, but finds he can describe what to look for only in negative terms, by describing what bone was not: "the brown concretion that wasn't quite brown, the texture that wasn't merely rock, the shape that couldn't be expected to have been bone but wasn't quite anything else." This is literally true, since you are in fact looking for something that is no longer there, the impression of bone in rock being much like the ghost of a moving object on a photographic plate. Then Dawe meets an itinerant photographer on the river, a certain Michael Sinnott, who has come West to collect photographs for his Travelling Emporium of the Vanished World, standing in the shallow river at Steveville saying, Everything vanishes. "We are two of a kind," he later tells Dawe. "Birds of a feather. You with your bones that are sometimes only mineral replacements of what the living bones were. Me, rescuing positive prints out of the smell of the darkroom." Dawe bridles at the analogy: "You make the world stand still," he retorts, "I try to make it live again." But the only real difference between them is that of verb tense: Sinnott records a world that is vanishing; Dawe floats through one that has long since vanished. Both are making a world come alive.

As we prospected, Don, too, was moving through a landscape that existed only as a ghostly image. The thin, black, coal seams near the tops of the coulees had once been the floors of ancient swamps; white bars meant beach sand; ripple patterns were the near-shore bottom of the Western Inland Sea. Don entered that world by reading the geology. At one point in our walk, when we were nearing the top of a gradual, striped inclined plane, he stopped and looked east: "We're standing up to our knees in swamp water," he said, "looking past a long, white-sand beach at row after row of slow, rolling waves." He didn't press it, but I was sure he was also seeing turtles, most of them small, painted plesiobaenids but occasionally a larger *Neurankylus*, clawing its arduous way from the sea to deposit its eggs in the warm sand, while thousands of long-legged shorebirds strutted about picking at crab shells. Many of the eggs would be dug up by oviraptors long before they hatched.

The turtle shell curved out of the side of a low hill like the edge of a thick, ceramic plate, one that had been sliced at regular intervals to prevent it from curling in the kiln. The hill was smooth, grey siltstone, crumbly as foundry sand to the touch. The shell was a hard disc; if left for another year or two it might have formed the top of a hoodoo. As it was, there were three feet of overburden to remove before we could touch the fossil itself, and we set to with our geology hammers and chisels, scraping away the siltstone a layer at a time. We worked at it for the rest of the morning, then halted for lunch. By then the heat was intense, perched as we were at the top of a small rise within a wider valley that caught and held the sun's heat like oven bricks. We were slowly being baked alive, happy martyrs to science. Just beyond the turtle hill, a sharply eroded ravine made a dark gash in the Badlands, angling down to a deep, dried-up stream bed, which in turn bifurcated into a labyrinth of canyons and cutbacks that radiated for mile after intricate mile. The maze of canyons was deep and shaded and looked cool. We sat drinking tepid water, talking about turtles, our eyes drawn to the shade, and after lunch we decided to give up on the turtle for a while and do some prospecting in the canyons.

The floor of the main canyon was a damp, braided streambed, littered with hadrosaur and buffalo bones, the extinct and the extirpated tumbled dizzily together by raging torrents of long-evaporated water. The walls rose perpendicularly, only occasionally broken by slumped indentations or sharp-edged tributaries. We stayed together for a while, then one by one drifted off to prospect alone in these cool, dark canyons, where even the air seemed to contain hidden whiffs of the past. I turned into an opening as narrow as a London alley, the walls so close I could reach out and touch both of them. The wall to my right was deeply undercut by a vanished stream; had I wished, I could have lain flat on the canyon floor and squirmed under it and been lost to sight. The wall itself seemed poised to collapse, lending the canyon a delicious element of danger. I turned a few corners, climbed over a few embedded boulders, and within minutes was deep in the labyrinth and completely alone. Everything else vanished.

I wasted some time digging around a half-buried thigh bone before realizing it was hadrosaur. Not worth taking out, I decided. I marked it with red surveyor's tape and moved on. Ten feet farther I found part of a vertebra stuck about eye level in the cliff. The centrum measured six inches across: somewhere deep in the canyon wall there might have been the rest of the skeleton, but I could see no place where it was coming out. Dave Eberth had told me about a prospecting trip he'd taken along the South Saskatchewan River, north of Medicine Hat, where in one summer he'd found fourteen separate bonebeds containing a total, he estimated, of ten thousand dinosaurs, all ceratopsians, all killed in a single, cataclysmic storm, possibly a hurricane. From the stratigraphical evidence, he calculated that the dinosaurs had died 124 miles in from the coast of the Inland Sea. "It is tempting to say that the animals migrated inland, not actually at the shore, where the storms were." Since it wasn't like Dave to give in to such rash speculation, I assumed he was speaking from a position of some authority. The storm must have been unusually violent to have penetrated so far inland and killed so many dinosaurs, the kind of storm, Dave said, that under ordinary circumstances occurs once every thousand years, or more often when the Earth is undergoing some major climatic change. These are the kinds of thoughts that spiral through your mind when you're prospecting: every bone will be significant, diagnostic, and connected to another bone, two bones together will lead to a complete skeleton, and a complete skeleton means a probable bonebed. Dave found it interesting that, although there were fourteen monospe-cific bonebeds in Dinosaur Park, no baby ceratopsians had ever been found in any of them; since babies hatched and remained near coastal areas until they were old enough to move inland to migrate, he'd said, it was unlikely that Dinosaur Park was coastal. It must have been inland, perhaps a higher plateau not unlike the Prairies today (this was me speculating now), which would have been ideal migrating terrain for giant herds of *Centrosauruses*, as it had been for giant herds of bison in modern times. Tens of thousands of dinosaurs annually trampling the length of North America, from the Arctic Circle to the Gulf of Mexico, drawn as though by instinct straight to the place where, ten million years later, the death comet would strike.

Perhaps all animals have an instinct for extinction, something coded into our DNA that lets us know when it's time to admit that we are no longer the fittest, when environmental change has accel-erated beyond our capacity for adaptation. When staving off the inevitable is more trouble than it's worth. When we might as well break out the champagne. Flowers anticipate death by bursting into bloom. Swans sing.

These thoughts got me so deeply lost in the maze of canyons that I decided I might as well continue walking until I came out somewhere. I had water and two granola bars in my field kit. Look-ing up, a hundred feet above my head was a narrow streak of blue sky between the black of the canyon walls; magpies and mourning doves soared determinedly across it. Up there, life was going on. Every now and then I passed a place where I could have climbed out to it, but I didn't want to. By now I was surely in a section that had not been explored in ages, perhaps not at all; something new and big was certain to be around the next bend.

When I came to a spot where the canyon met another, slightly wider canyon, the intersection opened up like a public square.

Suddenly there was sunlight and small bushes. To my left, at the very corner where, if this were a city and the cliffs were office towers, the newsstand and diner would have been, the cliff levelled off about six feet above the floor to form a triangular shelf that was feathered with bright green grass. I climbed up to it; there was even a handy log to sit on, though how it had got there I couldn't begin to guess. As I drank some water and ate a granola bar, I noticed something odd about the place, something slightly unnatural. The cliff behind me seemed to rise at an unnatural angle, and the scree at the bottom appeared to have been worked. I went over to look more closely and nearly stumbled over a large leg bone that was lying on the shelf, almost completely exposed. Beside it was another bone, perhaps a humerus or tibia. There were ribs, and smaller, broken bits of bone, and a centrum, lying flat like a drum on the sandstone: with a sudden thrill I realized I was standing on a dinosaur.

It was an old quarry. Many of the bones had been partially exposed and then left: others had eroded out of the cliff wall and lay loose on the floor. They appeared to belong to a hadrosaur, and there did not seem to be a skull. As I brushed sand from one of the ribs, I exposed a section of an old, faded newspaper. There was quite a lot of it; perhaps whoever had worked this quarry had used newspaper to pack the bones they took out into wooden crates. Carefully lifting the top page, I found I could make out part of a paragraph: "Washington AP," it began. "The US is sending a second fighter-bomber squadron to Germany, it was announced Wednesday. . . ." The article fixed the date of the quarry as sometime after Pearl Harbor, which was puzzling, since bone-hunting in the area was supposed to have been suspended during the Second World War. Once again, layers of history were intertwining in the Borgesian labyrinth of the Badlands.

CHAPTER THIRTEEN

Ghost Birds

P HIL AND EVA WERE sitting outside the kitchen trailer that
evening. They'd come down from Dry Island for a few days of
prospecting in Dinosaur Park, and had brought Rodolfo and Christian with them. We shook hands all around, and I started to greet
Christian in Spanish before realizing that I had forgotten almost all
the Spanish I'd learned in Argentina.

"*Como está?*" I mumbled, thinking I should have used the more
colloquial *Que tá?* I then tried to ask him how the elections went,
saying something with the word *elecciónes* in it, thinking at the
same time of Borges's book *Ficciónes*, and Christian said, "*Muy
bien.*" Rodolfo looked somewhat pityingly at me, and I decided to
switch to English.

"The old mayor is now the new mayor again," Rodolfo said.

"Is that good?"

"That's very good. He likes my museum. He has promised us a
new wing for it, as long as I put some of that old stuff from the oil
company in it. And Claudia's job is also secure. And we had a great
flight," he added. "We came through Los Angeles, and I stopped to
see Luis Chiappe. Now we want to see a hockey game."

"It's July," I said. "We don't play hockey in July." Although it
felt cold enough, at least in the mornings.

Rodolfo was disgusted. "What do you do, then?"

"Back home I play baseball. Here, I don't know. Break broncos, I guess." The Calgary Stampede was just about to begin. I looked dubiously at Phil. "Maybe we could take them to the Stampede?"

"Naw, forget it," said Rodolfo. "We can see gauchos any time we want."

I told Phil about the old quarry and the newspaper I'd found. He and Eva and Darren had been mapping old dinosaur sites for years: Charlie Sternberg's maps were far from complete, perhaps on occasion deliberately so; Sternberg would note that he had found a skeleton "ten miles south of Steveville," which could be anywhere, and Barnum Brown, as Jack Horner has noted, was equally "careless about collecting information," which might also be interpreted as careful about giving away collecting information. "He simply didn't consider it important to describe where he unearthed his speci-mens," writes Horner. Whatever their motives, Darren estimated that there were as many as two hundred unmarked quarries in the Park, and he and Phil were trying to fill in as many of the unknowns as possible. It was important to know which area yielded hadrosaurs, for example, and what geological level they came from; to fix them, in other words, in space and time. The early collectors were content to leave specimens in a kind of dimensionless limbo, shrouded in mystery, and the museums mounted exhibits that mod-ern paleontologists would call useless, because they said nothing about the real life of the specimen. Thoreau, an outdoorsman, sensed this: "What is it," he asked, "to be admitted to a Museum, compared with being shown some star's surface, some hard matter in its home?" To modern paleontologists, a specimen is only inter-esting in context, and often that context is measured in millime-tres. Darren used GPS equipment sensitive enough to mark a position's longitude and latitude to within a single centimetre. Phil wanted to re-open many of the old quarries for overlooked or mis-interpreted material. Brown and his contemporaries were mainly interested in trophy hunting, crowd-pleasing stuff: skulls, complete skeletons, or, as with Brown's *Albertosaurus* quarry, the feet, enough to identify a species and perhaps make a sale to an institution. Only complete skeletons were taken in their entirety, and even then the work was often hasty and much was missed. Charles Sternberg, who collected for the Museum of Nature, mapped a few sites in the Steveville area in the 1940s, but had not ventured this far inland. His brother Levi, who worked for the Royal Ontario Museum until the 1950s, opened many new sites, but his records were as vague as Charlie's had been.

"Speaking of trophies," I said, "how did the hadrosaur sacrum enjoy its helicopter ride?"

Phil groaned and Rodolfo got up to get another beer. Eva looked at me. "You mean you haven't heard?" she said.

"Heard what?"

I hadn't seen a newspaper or listened to a radio for weeks, but apparently the story had been hard-wired across the country for days, how the sacrum had been wrapped in burlap and plaster, girdled with two-by-fours, secured to the chopper's belly with cables. How the chopper had tried to rise, urging its cargo off the ground, a foot, then a yard, then out over the yawning mouth of the canyon. "It was too heavy," Phil said. "We estimated it at three hundred pounds, but it must have been heavier than that."

"It was probably more like six hundred," said Eva. "But the helicopter got it up over the valley."

"Then it started swinging," said Phil. "First the sacrum started to swing, then the helicopter. It was really scary. The pilot thought he was going to crash, and cut it loose."

"Was it smashed?" I asked.

"To smithereens," Phil said, shaking his head. "A million pieces. We watched it from the edge of the prairie, and when it hit the ground there was nothing but a puff of white plaster."

We sat silently for a while, thinking about the *Hypacrosaurus* pelvis being turned instantly to dust. It was more than wasted work, more even than lost data. It was a kind of sacrilege, as though a pall-bearer had slipped on the wet church steps in the rain and dropped the coffin.

There was, of course, another way to look at it. What if animals did have an instinct for extinction? What if the *Hypacrosaurus* had simply not wanted to be moved again? Enough, it might have said. It had travelled far enough, first chased by a pack of maurading theropods from its breeding ground, then carried down to a distant, tempestuous sea, subjected to the slow, the infinitely intimate drip of fossilizing water. Let me rest, it might have said. Not even rock, let me return to dust.

"History," André Aciman writes, "no matter how often it comes back, is always about rubble and the piling up of stones. It aches for extinction."

MY CAMPSITE BESIDE the Red Deer River was exactly where Barnum Brown had tied up his barge and made camp in 1912, after floating downriver from Drumheller, past Steveville and the worst of the Badlands to the roadhead to Patricia. There was a break in the bank just there, where I could walk down to the water's edge and watch the bank swallows skim the river's sinuous surface. After the rain farther north, the river was swollen and swift, thick with sediment and laden with debris. Branches floated by, waving their leaves in the air like shiny green flags before rolling over to be buried in the mud-brown water. Huge chunks of clay, some as large as corpses, bobbed and nodded past, surrounded by billows of brown foam. Occasionally, one of them would sink, and the river would make a sort of satisfied, sucking sound, like the licking of lips.

Human beings have always lived beside rivers, so that in our imaginations life, rivers and time have become one. Rivers move, motion is time, and time enmeshes our lives like a cortex. According to the authors of Genesis, "a river went out of Eden to water the garden, and from thence it was parted into four heads." The rivers linked Eden to the rest of the world: the first, Pison, flowed into the land of Havilah, which was probably southern Arabia or Persia, from whose language the Greeks took the word "paradise." The second river was the Gihon, which flowed into Ethiopia, the original homeland of the Egyptians who, descendants of the sons of Ham, were almost certainly black. The third and fourth rivers were the Tigris and the Euphrates, the alpha and the omega, the twin rockers on the cradle of human life. By identifying the four rivers, scholars located the Garden of Eden, at least spacially. But the rivers also defined Eden temporally: they were not *wádi*, torrential but brief streams that, like the arroyos of the Patagonian steppes, roared with water after the rains but were reduced to highways of hot rocks the rest of the year. Eden was fed by true rivers, *naharim*, permanent, perennial, immortal.

From the green marsh across the river a reed detached itself from the rest, uttered a dispirited *Gronk!* and took off, doubling its long, jointed neck under its beak and slowly flapping its huge wings: a great blue heron. Of all the birds I had seen, shorebirds seemed to me the most reptilian; their ageless, yellow eye, their reluctant flight. Even their preference for fish, the most ancient of foods, betrayed their lineage, like an old man eating olives. The heron flew low to the water, so close that I could see the ripples made by the rush of air from its wings. In *Badlands*, Kroetsch describes herons that "did not move into the air but moved out of the water," which also describes their evolution. Then my heron hauled itself with much effort towards a cottonwood tree, and landed "awkwardly," as Kroetsch writes, its "forgotten legs remembered only just in time."

Of course, a thing can only be remembered in time. A friend who has died lives still in memory, since all time, past, present and future, co-exist and, as Norman Maclean observed, "a river runs through it." One of the books in Borges's permanent library was J. W. Dunne's *An Experiment With Time*, in which Dunne, according to Borges, "states that the future, with its details and vicissitudes, already exists." Life flows towards the pre-existent future on "the absolute river of cosmic time, or the mortal river of our lives." A person standing on the bank of this river, observing its passing, taking note of the clods of earth floating in it, or of the great blue heron floating through the air above it, can be aware that he is doing this only if he is also outside himself observing himself observing the river, in which case that second self is also being observed by a third self, and so on to infinity. Borges calls this the "vertiginous mystery."

WILLIAM CUTLER WAS responsible for many of the unmarked quarries in the Park, and also camped here on the river. Cutler was a local farmer and collector who became so entirely obsessed with fossils, so avid for them, that it killed him. He was born in England in 1878, emigrated to the United States, fossil hunted in Wyoming, turned up in Alberta in 1910, and two years later presented himself to Barnum Brown, perhaps on this very spot on the river bank, with an offer: he would guide Brown to as many fossil sites in the area as he knew, and in return Brown would teach him how to collect them. To seal the partnership, Cutler gave Brown a skeleton he had already taken out, a single-horned ceratopsian that Brown named *Monoclonius cutleri*. When Brown left, Cutler continued to collect: in 1915 he sold an ankylosaur to the British Museum,[1] where the famous paleontologist Franz Nopsca examined and named it

---

1 W. P. Pycraft, in his history of the British Museum of Natural History written about this time, writes that "the study of fossils is like the study of fragments of some old books in a long-forgotten language; only bits here and there of the earth's history can be traced by these broken and battered remains, but enough can be made out to furnish a most wonderful story."

Scolosaurus cutleri. One winter, he was found huddled in a tent in the Badlands, suffering from pneumonia and very nearly dead. Shortly after this, in 1922, he parlayed his field experience into a teaching position at the University of Manitoba; to get there, he bought a row-boat in Edmonton and rowed it down the North Saskatchewan River, all the way to Winnipeg.

His exploits so impressed the British Museum that in 1924 he was hired to lead its East Africa Dinosaur Expedition into Tanzania. And now two bizarre strands of history briefly intertwine. The previous year a young Cambridge student named Louis Leakey, who would later discover in Africa the hominid remains that now define us as a species, received two kicks to the head in a single rugby match, suffered a bout of post-traumatic epilepsy from them, and was ordered by his doctors to take a year off from his studies to do something relaxing, like, oh, let's say, fossil collecting. Leakey had been born in Kenya, knew some African languages and figured he could stand the heat, and signed up for the British Museum expedition as Cutler's chief, in fact only, assistant and interpreter. "I little thought when I was kicked in the head," he later wrote, "what a great effect that incident would have on my career."

Africa would not be an immediate success. Tanzania had belonged to Germany before the Great War, and now it belonged to Britain. When in German hands, a tremendous fossil site at Tendaguru had yielded a wealth of marine creatures—corals, bivalves, cephalopods, gastropods—as well as five dinosaur families, including the gigantic sauropod *Brachiosaurus*. In 1911, for example, Werner Janensach and Edwin Hennig had gone in with 480 local labourers and worked more than one hundred sites over thirty square miles of territory, and removed half a million pounds of fossils; Dale Russell, who tried to re-open the sauropod site in 1977, compared it to the famous Morrison Formation in the western U.S., which had held *Diplodocus* and *Apatosaurus* (a.k.a. *Brontosaurus*), as well as numerous fish, frogs, lizards and pterosaurs. The British Museum wanted its own *Brachiosaurus*, and sent Cutler and Leakey to procure one.

In 1924 Tendaguru was a miasma of infectious insects and pointless official inertia. It didn't help that Cutler and Leakey loathed each other on sight, the self-taught loner and the inexperienced scholar, two colonials anxious to impress the mother country. Leakey described Cutler as a man of "little eccentricities and pretensions." Cutler viewed Leakey as an annoying necessity. When they arrived in Dar es Salaam, Cutler stayed behind to secure papers while Leakey continued on to Lindi, where he would hire the workers. Cutler would join them in Tendaguru when everything was set up. He stayed away two months. When he finally arrived, he spent more time collecting butterflies than doing the hard work in the quarry, and then he wouldn't let Leakey touch an unjacketed specimen. He had learned the art of burlap and plaster from the great Barnum Brown, and no mere pipsqueak of an etc., etc. Leakey appears not to have minded, probably because he had come down with malaria and was running a fever of 104. (Cutler's journal is descriptive—"Leakey passing blood and vomiting"—but you can almost hear him telling the younger man, "Why, this is nothing! I remember one winter in the Alberta Badlands...") He had the workers digging trenches in the swamp, fifty feet long and twenty-four wide, but the sauropod bones in them were poorly preserved and badly eroded, as sauropod bones often are.[2] Eventually, Leakey left, the combination of malaria and epilepsy doubtless proving too much for him. Cutler stayed valiantly on for another nine months, until he, too, succumbed to malaria. He died, in camp, on August 30, 1925, by which time the

---

[2] Many of the sauropod bones we found in China were little more than uncollectable smears of white on the ground, and conditions there were relatively favourable.

British Museum seemed to have lost interest in the venture: his replacement was F. W. H. Migeod, who is known to history only as "an African traveller."

I like to think of Cutler as he was in Canada, rowing his boat along the North Saskatchewan to take up his post in Winnipeg, where his students would call him Old Cuttlefish. I see him pulling into his future as he looks back at his past, or scanning the cliffs beside him for glints of bone, tying up from time to time at a snag and wading ashore to make tea, or to examine a promising outcrop. This was the life, he must have thought.

STEPHEN JAY GOULD, on the other hand, does not think of life as a river. "But life is not a tale of progress," he writes; "it is, rather, a story of intricate branching and wandering with momentary survivors adapting to changing local environments, not approaching cosmic or engineering perfection."

Life, in other words, doesn't rise like a phoenix; it spreads like a cancer.

SOMETIMES, WITH OUR help, it sinks like a stone. The great auk, the passenger pigeon, the Labrador duck, the dodo. We now say "Dead as a doornail," but the original expression was "Dead as a dodo." You can't get any deader than extinction. The dodo disappeared in 1641, a few years after it had been seen by English mariners on a small group of islands off Madagascar. "The progress of Man in civilization," wrote H. E. Strickland and A. G. Melville in *The Dodo and Its Kindred*, published in 1841, "no less than his numerical increase, continually extends the geographical domain of Art by trenching on the territories of Nature." Yes, we have seen this in our time. "Hence the Zoologist or Botanist of future ages will have a much narrower field for his researches. It is, therefore, the duty of the naturalist to preserve to the stores of Science the knowledge of these extinct or expiring organisms, when he is unable to preserve their lives."

The burrowing owl, Kirtland's warbler, trumpeter swans, the California condor, the Eskimo curlew.

IN PATAGONIA, DURING one of our breaks from the quarry, I descended alone to the dry riverbed and lay down on the sand. Lying on my belly, I faced upstream, like a fish, from which perspective the riverbed seemed as wide and flat as the world was meant to be. Grass grew in some of the older channels and was also lying down, pressed to the earth by the force of the recent rain. Hearing a soft snicker behind me, I turned and saw two horses making their way up the canyon, a stallion and a mare. I'd known there were wild horses in the valley—the riverbed was littered with their bones and footprints—but I'd never seen them. These looked old and tired. The stallion walked slowly, his head sagging almost to the ground, as though he were afraid of missing a single blade of grass or trickle of water. I thought he might have been blind. I thought of a word that applied both to the horse and to the river, and later wrote it in my notebook: "cataracts." The mare stared incuriously at me as she passed.

THE BADLANDS BEGAN a short hike inland from my campsite, and in the evenings after dinner, while the others played Frisbee or read in their trailers before going down to work in the field station, I would walk to the nearest coulees with my binoculars and collecting tools. Away from the river the landscape turned grey and ashen, but there was furtive life in the cool hills when the insects came out and birds swooped down on them from above. They were nighthawks, known in Europe as goatsuckers for their large mouths (for which they are also called nightjars) and their suspicious habit of hanging around barnyards. I preferred to call them nighthawks;

they were black and thin of wing, with a bright white spot at each wrist. In the enveloping twilight, they moved like ghosts above the luminous clay.

In South America one of the nighthawk's numerous names is *urutau*, which means "ghost bird." In his essay "Goatsucker Myths in South America," Claude Lévi-Strauss points out that among the Kalini people of Guyana, human beings have several souls, one of which remains on Earth in the form of a nighthawk; a bird thus links the living with the dead. There are also stories in which a rejected lover, usually a lower-class woman whose affair with a nobleman is prohibited by her lover's family, retreats to the forest and changes into a nighthawk: she dies but remains among mortals, like a ghost, and brings messages to them from the gods or from the netherworld. Elsewhere, because nighthawks were often seen in graveyards, they were considered omens of death and were never hunted: among the Tehuelche of southern Argentina, the giant race Magellan first called Patagons, nighthawks were evil spirits that pursued humans, bit them with their wide beaks, and sometimes killed them. On the other hand, Jean de Léry, a sixteenth-century traveller, reports a version that I prefer to believe, that the Tupi Indians of the Amazon basin paid particular attention to nighthawks because, writes Lévi-Strauss, "the birds bore good news and encouragement from dead friends and relatives."

Aesop wrote that the first living creature on Earth was a bird; Aristophanes made fun of the idea in his play *The Birds*, performed in Athens in 414 B.C., in which the bird kingdom decides to regain its former glory by building a walled city in the air, to be called Cloudcuckooland, from which they would intercept smoke from sacrificial fires on Earth before it reached the gods, and also charge a tariff to any gods passing through the city on their way to rape and plunder the Earth. They would control everything. "Listen, you men down there in the half-light!" they would say to us, "shadowy, impalpable, dreamlike phantoms: feeble, wingless, ephemeral creatures of clay, dragging out your painful lives till you wither like the leaves and crumble again to dust! Pay attention to us, the immortals; to us, the eternal, the airborne, the un-ageing, the imperishable; and hear from us the whole truth about what lies around and above you!"

But the scheme fails, of course, and we never hear what the birds had to say.

In England the nighthawk is the poor-will, Lévi-Strauss says in imitation of its cry: "Five notes, three of which are not heard."

As I walked through the gathering twilight I continued, out of habit, to scan the bases of the cliffs for signs of bone. Although this close to the Park's permanent base camp the cliffs must have been prospected hundreds of times, the recent rain might have brought down a fresh scree of material. Volcanic clay slides downhill when it's wet, which is why the Badlands are renewed, always giving up more dead. At the bottom of a small, cone-shaped hill, I came upon a few fragments of eggshell and stopped to pick them up. They were modern: from their size and shape I guessed that they came from a hawk or turkey vulture egg, and thought the nest must have been at the top of the hill. I circled the base, found a way up, and climbed using my geology hammer as a pick. I kept finding more bits of shell. At the top, there was indeed a slight concave depression, and in it I found the remains of at least one more egg. But the startling thing, the numinous thing, was the depression containing the eggs. The bird's chosen nest was the partially exposed skeleton of a dinosaur.